Mohammed Kouidri
Mohamed Amine Nedjma

The golden aphid of the Laghouat Atlas pistachio tree

Mohammed Kouidri
Mohamed Amine Nedjma

The golden aphid of the Laghouat Atlas pistachio tree

Evaluation of the effects of golden aphid

ScienciaScripts

Imprint

Any brand names and product names mentioned in this book are subject to trademark, brand or patent protection and are trademarks or registered trademarks of their respective holders. The use of brand names, product names, common names, trade names, product descriptions etc. even without a particular marking in this work is in no way to be construed to mean that such names may be regarded as unrestricted in respect of trademark and brand protection legislation and could thus be used by anyone.

Cover image: www.ingimage.com

This book is a translation from the original published under ISBN 978-620-6-71277-0.

Publisher:
Sciencia Scripts
is a trademark of
Dodo Books Indian Ocean Ltd. and OmniScriptum S.R.L publishing group

120 High Road, East Finchley, London, N2 9ED, United Kingdom
Str. Armeneasca 28/1, office 1, Chisinau MD-2012, Republic of Moldova, Europe
Printed at: see last page
ISBN: 978-620-7-66418-4

TABLE OF CONTENTS

INTRODUCTION

The genus, Pistacia, belongs to the order Saphindales and the family Anacardiaceae, and is of Asian or Mediterranean origin (Karimi et al., 2009). The Atlas Pistachio (Pistacia atlantica) is a species that extends from south-west Asia to north-west Africa (Zohary, 1952). In Algeria, the Pistachier de l'Atlas grows wild in sub-humid areas as well as in Saharan regions (Kadi-bennane et al., 2005), and its extreme limit is in the Hoggar (Monjauze, 1980; Seigne, 1985; Al-saghir, 2010). It is concentrated on the Arbaa plateau in southern Algiers, hence the name dayas country (Monjauze, 1980). This forest species adapts to all soils except sand, and is content with low rainfall of around 150 mm and sometimes less (Benhessaini and Belkhodja, 2004). It grows very slowly, but has the advantage of being the only tree capable of organising pre-forest ecosystems in arid and semi-arid bioclimates. In Algeria, its cultivation remains low despite its potential to adapt to arid conditions, and most of the country's mountainous agricultural regions are favourable to its spread. Bétoum generally develops in a sparse and isolated form, and is subject to very strong biotic and abiotic pressures which greatly limit its expansion and development (Monjauze, 1980). It is in a precarious and alarming situation because of its advanced degradation; the excessive use of its wood, pests, diseases and drought have all contributed to its degradation (Benabid, 1986; Benhssaini and Belkhodja, 2004). Unfortunately, very few studies have looked into these aspects, and even fewer have revealed the diseases and pests affecting this species (Belhadj et al, 2008; Mansour, 2011; Martinez, 2008, 2009). These aphids are among the most important plant pests, and over the course of their evolution they have developed remarkable abilities to adapt to the environment, with high fecundity, varied reproduction methods, alternation of winged and apterous individuals and use of several types of plant (Hulle et al., 1998). The damage Some attack the roots, but most live on leaves, stems, buds, flowers or even fruit. Our work consists of identifying and assessing aphid damage to stands of Atlas Pistachio in three sites in the Laghouat region. The aim is to characterise these stands from a biometric point of view, identify and quantify the presence of different aphid species, while specifying the role of the golden aphid and trying to establish a link between the biometric parameters of the Pistachio tree and the damage caused by this aphid species.

PART I
BIBLIOGRAPHICAL SUMMARY

CHAPTER 1
DAYA AND ATLAS PISTACHIO

Pistachio trees are characteristic of the Mediterranean region (Boudy, 1952). Throughout the world, steppe trees such as the Atlas pistachio (Pistacia atlantica Desf.) in arid bioclimates and its state of degradation require immediate and effective attention (Benhassain et al., 2004).

1. Les Dayas

1.1. Definition of Daya :

Dayas are small depressions which, in the endoreic regions of the Maghreb, dot the rigid surface of plateaux protected by lacustrine limestone or a pedological crust (Nesson, 1967). These are closed depressions where run-off water, or even small wadis, collect. This water stagnates and remains for a long time thanks to the low-permeability clayey silts that line the bottom of the depressions (Despois, 1949). In contrast to the zahrez and sebkhas (improperly called chotts), where the water and soil are highly saline, the dayas contain fresh water that is favourable to vegetation. These characteristics explain why the dayas are occupied by herbaceous and arborescent vegetation, whose economic role is far from negligible on these steppe plateaux (Estorges, 1961). These generally unmarked depressions are marked by the haughty bearing of old pistachio trees (Pistacia atlantica, in Arabic: Bétoum), some of which are several hundred years old (Agabi , 1995).The shrub layer is represented by jujube trees (Zizyphus lotus), which often form impenetrable thickets; finally, the herbaceous layer is a popular pasture for flocks of sheep. The dayas are also game reserves. In the last century, gazelles and ostriches benefited from the shelter and food reserves provided by the dayas; today, only bustards, hares, partridges and gangas can be found in the dayas. still hunted. The ostrich was hunted down and disappeared in the second half of the 19th century (Agabi, 1995).

1.2. Origin of dayas

The origin of the dayas has interested geomorphologists. Since a famous study by Capot-Rey in 1937, it has been accepted that these depressions, which are very similar to sinkholes, are of karstic origin. They result from the subsidence of the surface crust or the upper limestone layers following the dissolution of the underlying sediments by seeping water. Potholes are rare but not totally absent (Agabi, 1995).More recent studies, in particular those by Estorges in 1959 and

1961, have shown that karstic erosion is not the only cause of the evolution of the dayas. There are reports of dayas crossed by a perfectly defined wadi bed; there are also real strings of dayas connected to each other by stretches of wadi. Finally, in the Larbaa plateau, which is the typical region for dayas, to the south of Laghouat, there are extensive areas where no karstic phenomena occur. In short, Estorges felt that while there was no doubt about the original karstic erosion, run-off and mechanical erosion had done more than just alter the shape of the dayas in detail; their action appears to have often been decisive (Estorges, 1961).

1.3. The dayas of the arid and semi-arid region of Algeria

Dayas are particularly numerous in the Lower Sahara region, which is bounded to the south by the Mzab chebka, to the east by the oued Righ, to the north by the oued Djedi and to the west by the oued Zergoum. This region is traditionally known as the Land of the Dayas, and is also known as the Larbaa Plateau or the Arbaa Plateau. There are, of course, dayas outside this region and their names are often used in toponymy, but nowhere are they so numerous and play such an important role in people's lives (Taïbi, 1999). The Larbaa have their grazing lands on the plateau, but they are increasingly sowing cereals in the dayas, to the detriment of the pistachio trees, which have long since stopped reproducing. In the north of the region and along the Oued Djedi, other nomads come to winter: the Ouled Naïl from the Saharan Atlas, who also grow cereals. cereals in the dayas and wadi beds. Some dayas, which have long been cultivated, have become islands of sedentary settlement in the middle of the steppe zone and have given their name to the agglomerations that have sprung up in contact with the cultivated land (Nesson, 1967).

1.4. Morphological stages of dayas :

Generally speaking, the youngest are small (metric to decametric), round and not very deep. The older ones, which are relatively large (kilometre-sized) and irregular in shape, are limited by steep slopes that can be several metres high and cut into the limestone crust covering the hamadas (Taïbi, 1999).
According to Taïbi (1999), five stages of morphological evolution correlated with the evolution of the vegetation have been distinguished and highlighted by satellite remote sensing. Run-off, wind deflation and dissolution processes have given the dayas their present configuration and are at the origin of the vegetation that colonises them (Taïbi, 1999).

a) First stage

During the first stage, the dayas, which are small in size (averaging less than 30,000m^2), are characterised by great regularity of shape: they are almost perfectly circular. Their central zone, flooded for longer than the edges and clogged by a sandy-loam colluvial formation, is colonised by more or less dense vegetation, herbaceous (perennial species) and shrubby remeth (Haloxylon scoparium); it is surrounded by an aureole, drying out more and more rapidly towards the outer edges and subject to slightly longer deflation than the central zone, which is characterised by sparse vegetation then in transition with the surrounding steppe (Fig. 1). The subdivision of this class into 2 (classes 1 a and 1 b) corresponds solely to a difference in size (Taïbi, 1999).

b) Second stage

In a second stage, as the daya deepens, the jujube tree (Zizyphus lotus) gradually eliminates the previous plant association (Fig. 1 b). These dayas, with an average size of less than 100,000 m^2 , still have shapes close to the circle (Taïbi, 1999).

c) Third stage:

The next stage is the tree layer. Betoum (Pistacia atlantica) develops in the shelter of the jujube bushes, the herbaceous vegetation being pushed to the extreme periphery of the daya (Fig. 1 c). Generally speaking, the centre of the dayas is then covered by dense plant formations with trees, surrounded by increasingly loose herbaceous and shrubby vegetation (Taïbi, 1999).

d) Fourth stage:

The dayas with a longer morphological evolution show an even different concentric organisation: the centre is bare, with more or less dense vegetation confined to the periphery, indicating a long evolution which culminates in the drying out of the central zone by gradually eliminating all vegetation (Fig. 1d). At the extreme periphery, transitional vegetation towards steppe appears.

These large dayas (average surface area 170,000 m2) correspond to an advanced stage of evolution, for which intermediate stages can be defined. Eventually, the bare central zone extends until the vegetation completely disappears. The daya is then dead, the central zone completely bare or colonised by a fairly loose steppe of esparto (Stipa tenacissima) and spartum (Lygeum spartum) is surrounded by an aureole of very sparse low vegetation (Taïbi, 1999).

e) Fifth stage

At this stage, the dayas are the largest (average surface area over 300,000 m^2), have contoured shapes and are bordered by cliffs. However, their morphological evolution does not always follow the same processes as those of the other classes. Their size is not necessarily representative of a longer evolution than the others, but often of the exploitation of a pre-existing talweg (Taïbi, 1999).

Figure 1: Morphology and vegetation of dayas from incipient to adult stage

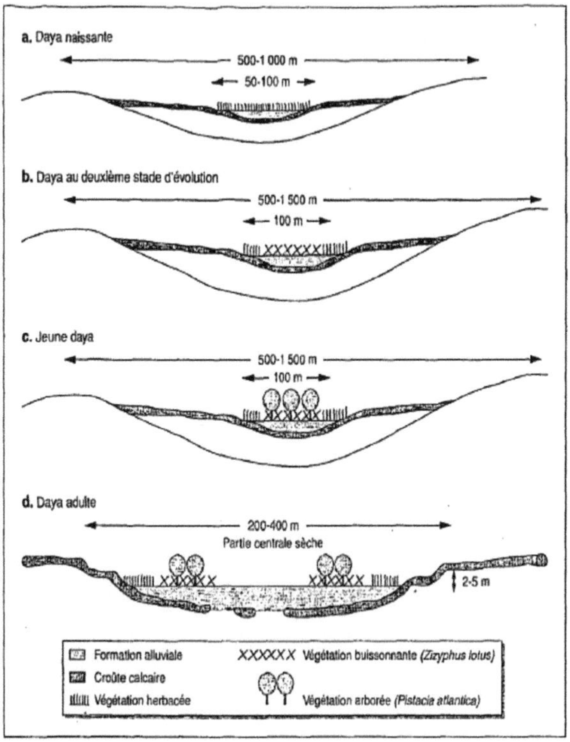

Source: (Taïbi, 1999).

2. Atlas Pistachio (Pistacia atlantica Desf.)

The Atlas Pistachio (Pistacia atlantica Desf.) is a tree whose main distribution area is in North Africa (Morocco, Algeria and Tunisia). It is also found in the Canary Islands, Libya (Cyrenaica), Cyprus and the Near East (Quézel et Médail, 2003).

The Atlas pistachio is a fairly common species in Algeria, but it thrives in arid and semi-arid regions, particularly the Hautes-Plaines1 where it thrives in wadi beds and dayas (Harfouche et al, 2005).

2.1. Botanical characteristics of the pistachio tree from l'Atlas

The Atlas key pistachio is a deciduous tree that can reach heights of 15 to 25 metres. It is known for its long life: trees with a circumference of 2.5 m are around 200 years old, and some of the oldest are as old as 300 years (Monjauze, 1968). They can reach 300 years (Monjauze, 1968). The bark is first red, then greyish, fairly light, then becomes a hard, fissured rhytidome. The foliage is ball-shaped when young, becoming hemispherical later on. Its roots are highly taprooted, reaching depths of 5 m (Monjauze, 1980).

2.2. Systematic :

Botanical classification of Pistacia atlantica Desf. (Yaaqobi et al., 2009) :

Kingdom: Plantae **Phylum**: Tracheobionta **Super-division**: Spennarophyra
Division: Magtioliophyra
Class: Magnoliopsida **Subclass**: Rosidae **Order**: Sapindoles **Family**:
Anacardiaceae **Genus**: Pistacia
Species: Pistacia atlantica

2.3.Pistachio distribution area l'Atlas

2.3.1. Range of the Atlas Pistachio in the world

Pistacia atlantica is a common Mediterranean species in Berberia, and is also found in the Middle East, in the Syrian desert and steppe and in Iran (Boudy, 1955). It is also found in the Crimea and Afghanistan (Seigue, 1985). Somon (1987) notes that the Atlas pistachio is a tree native to North Africa. Some authors are unanimous in stating that it is endemic to North Africa, where it is found in the northern Sahara (Fig. 2), in the Dayas at the foot of the Algerian and Moroccan Saharan Atlas (Quézel et Santa, 1963; Ozenda, 1991).

Figure 2: Natural range of Pistacia atlantica (Al-Saghir, 2006)

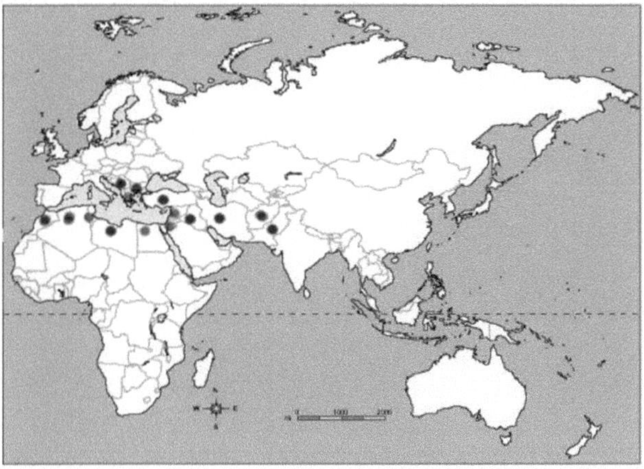

2.3.2. Distribution area of the Atlas pistachio in Algeria

It is an endemic species and one of the protected non-cultivated plants in Algeria (Kaabeche et al., 2005). According to Boudy (1952), in Algeria it is found (Fig. 3) scattered in the warm forests of the southern tell, but above all in the steppe-desert region of the high plateaux and the northern Sahara, where it is only found in the following areas in the Dayas. It is sometimes found in the mountains of the Saharan Atlas (Ain Sefra region) and on the high plateaux of Oran. Bétoum is a tree par excellence of the daya of the southern foothills of the Saharan Atlas, its extreme limit being in the heart of the Hoggar where it exists as a relic (Manjauze, 1980). It is mainly found in the transition zone between the steppe and the tell.

Figure 3: Distribution of Pistacia atlantica in Algeria (Monjauze, 1968)

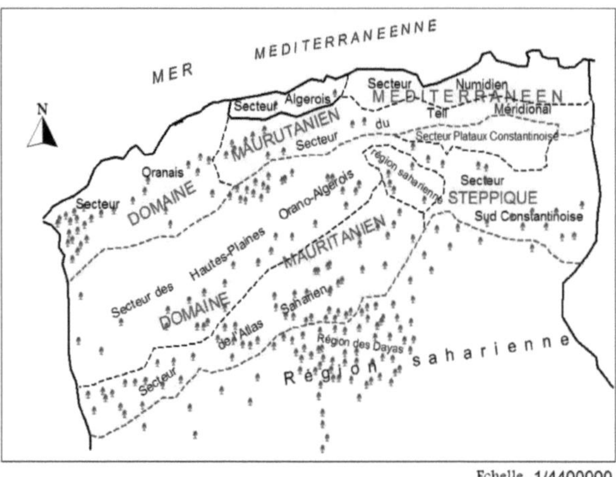

2.4. Morphological description :

The genus Pistacia, which belongs to the Anacardiaceae family, includes numerous species that are widespread in the Mediterranean and Middle Eastern regions. The Atlas pistachio tree (Pistacia atlantica Desf.), commonly known as El Betoum or Botma in Arabic, is a woody, spontaneous species that can reach 15 m in height. The tree has an individualised trunk with hemispherical foliage (Quézel et Santa, 1963). Its compound leaves consist of seven to nine leaflets, the flowers are in loose clusters and the fruits are pea-sized drupes (Ozenda, 1983).

2.4.1. Leaves :

Deciduous, semi-evergreen alternating with a finely winged rachis, irregularly imparipinnate with 5 to 11 odd-numbered leaflets, the pairs numbering 3 to 4 whole, oblong lanceolate (2.5 to 5 × 1 to 1.5 cm), obtuse at the apex, sessile and glabrous (Somon, 1987), they are about 8 to 20 cm long (Boulos, 2000). The petiole is not winged and is 3 to 5 cm long. The rachis is flattened and barely winged (Zohary, 1987). Their colour varies from dark green on the upper surface to light green on the lower surface (Khaldi and Khouja, 1995). They are slightly leathery and rarely measure more than 12 cm in total length, their greatest width being in the lower third of the leaf blade (Photo 1). In autumn, they blush opportunely in gardens (Manjauze, 1980).

Photo 01: Atlas Pistachio leaf in the study area (early October, Mrigha 2017)

2.4.2. The inflorescence :

The Atlas pistachio has an inflorescence in the form of a raceme. Flowering begins in mid-March, just before foliage sets (Yaaqobi et al., 2009).

2.4.3. Flowers :

The male and female flowers are borne on different plants. However, a few monoecious plants have been observed whose male and female flowers are borne on different branches. No hermaphroditism has been observed. The flowers are small in axillary panicles and are apetalous. They are regular flowers with a tendency towards zygomorphy (Yaaqobi et al., 2009).

2.4.4. The fruit :

The fruit is a drupe, whose vernacular name is "Khodiri". It is eaten by local people (Belhadj et al., 2008). Fruiting begins towards the end of March and the fruit reaches maturity in September (Yaaqobi et al., 2009).

2.4.5. The bark

The bark has longitudinal cracks (Khaldi and Khouja, 1995), and produces a putty resin that naturally exudes abundantly in hot weather (Belhadj, 1999).

2.4.6. Other characters

a) Growth :

According to Quézel and Medail (2003), growth is very slow in the wild, but in irrigated plantations it is quite rapid (30cm/year, sometimes more). It can live for around 300 years.

b) Regeneration :

Natural regeneration of Betum remains very uncertain and difficult, mainly because of the hardness of the integuments, which inhibits germination. It rejects stumps well (Boudy, 1952). According to Riedacker (1993), the germination rate in nurseries hardly exceeds 20%.

c) Pollination :

Only the flowers of male plants attract bees, which actively collect pollen. However, they do not play any role in pollination because the female flowers are not visited. Pollination remains solely anemophilous (Yaaqobi et al., 2009).

d) Entomology :

These include the golden aphid, which causes blisters or galls on the leaves (Belhadj, 1999) and is sensitive to Verticillium dahliae (Monastra et al., 2005).

2.5. Ecological requirements

It is one of the rare tree species still found in semi-arid, arid and even Saharan regions. This exceptional plasticity in relation to atmospheric drought could be its main characteristic, but it is no less indifferent to the nature of the soil and can occupy the most extreme situations within its botanical range (Manjauze, 1980). It is a main species that is currently found in a disseminated state, which adapts to arid climatic conditions and can live in the most severe ecological conditions (Boudy, 1952).

2.5.1. Requirements climate

a) Rainfall :

This species does not require much rainfall, as it is found in the Mitidja with rainfall exceeding 1000 mm per year and in the south at Ghardaïa with 70 mm per year (Manjauze, 1980).

b) Temperature

According to Larouci and Ruibat (1987), the Atlas pistachio is a heliophilous species and is resistant to both low and high temperatures. It can grow from -1 2°C to over 49°C (Kaska, 1994).

2.5.2. Requirements

Indifferent to soil type (Zohary, 1996), Betum is very undemanding from an edaphic point of view, adapting to a wide range of soils: from acid silica soils to calcareous soils in Syria, with the exception of sandy soils (Boudy, 1955). Clay soils and lowland alluvial deposits: It is found fairly rarely on limestone rock in dry mountains, it is confined to depressions (Boudy, 1952). The species grows well in clay or silty soils, although it can also develop on limestone rocks (Khaldi and Khouja, 1996).

2.5.3. Altitude

According to Boudy (1952) and Monjauze (1968), this tree grows best between 600 and 1200m. It can reach an altitude of 2000m in the dry mountains. And according to Zohary (1952) up to 3000m to the east of its range. According to Belhadj et al (2008), the species is found at 107m (Guerrara station).

2.6.Interests

2.6.1. Agro- ecological value

It is used for reforestation in the most severe sites to combat desertification. It also plays a role in soil conservation and is used as a windbreak to fix dunes. It is an excellent rootstock for the true pistachio tree, which is more resistant to root asphyxia than other species of the Pistacia genus. It is a source of energy, using its wood for cooking and heating in regions where living conditions are particularly poor. It is a source of shade: animals find P. atlantica a good refuge from the sun's heat and radiation. The tree is often the only tree in the region (Chaba et al., 1991).

2.6.2. Interest economic

Its interest is as follows: Rootstock for Pistacia vera, because of its resistance to aridity, its overly powerful root system and its low climatic requirements (Daneshard et al., 1980 in Maamri, 2008).Local people living near populations of Pistacia atlantica Desf. use these fruits as food and provide an edible oil.

(Chaba et al., 1991). This oil is extracted from the seeds, which contain around 55% (Daneshard et al., 1980 in Maamri, 2008).The Atlas pistachio is a reforestation species, with around 100 hectares reforested each year as part of the Green Dam (Chaba et al., 1991).

2.6.3. Benefits medicinal

Production of oil with high nutritional value: oil extracted from the seeds presents interesting prospects. The drupes of the Atlas pistachio have a very appreciable oil yield of around 40٪, compared with those of other species such as soya (20 to 22٪), olive (20 to 25٪). Analysis of this oil has highlighted its composition in various biochemical constituents such as: carbohydrate structures (statured fatty acids and unsaturated fatty acids), sterols and various vitamins (A and E). The bark produces a mastic resin. Local people use it for medicinal purposes. The leaves and bark are used as a decoction to treat stomach aches and pains. When inhaled, the leaves are used as a febrifuge. The galls are used in powder form, alone or in combination with yellow nutsedge, as an antidiarrhoeal and stomachic agent (Lamnaouer, 2002).

2.6.4. Fodder value

Pistacia atlantica is a valuable species because of the various benefits its leaves can provide. In times of food shortage, the tree can supply livestock with up to 0.35 fodder units.

CHAPTER 2
APHIDS

The super-family Aphidoidea comprises around 4,000 species of insect in the order Hemiptera, divided into ten families. Of these species, around 250 are agricultural or forest pests, generally known as "aphids". ". They vary in size from one to ten millimetres in length (Fraval, 2006).

2.1. Systematics

According to Iluz (2011), aphids are classified as follows

Kingdom: Animalia **Phyllum**: Arthropoda **Class**: Insecta
Order : Hemiptera

Suborder: Stemorrhyncha **Super family**: Aphidoidea **Family**:Aphididae
Adelgidae Eriosomatidae Phylloxeridae
According to the same author, the suborder Stemorrhyncha also includes other insects such as psyllids, whiteflies and scale insects.

2.2. Description

Aphids are small insects measuring between 1 and 4 mm that live in colonies on many crops and weeds. All individuals have a pair of generally well-developed cornicles on the back of the abdomen, a cauda (tail at the end of the abdomen) of varying length, antennae and a frontal sinus (Fig. 4). All these organs are characteristic of each species (Leclant, 1999).

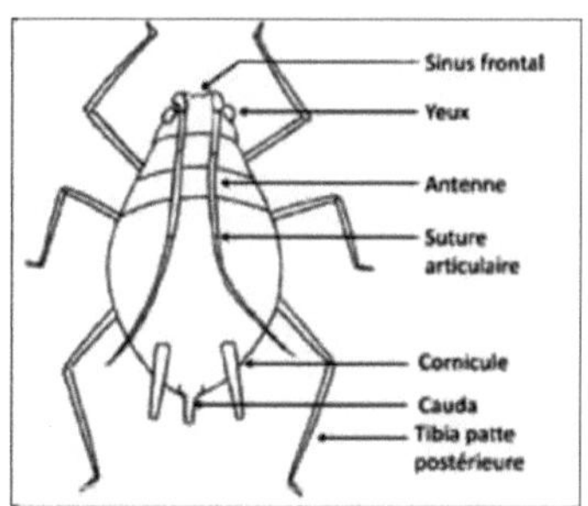

Figure 4: Diagram of an aphid (Godin and Boivin, 2000)

Aphids have a rostrum (a biting, sucking stylet) on the ventral side of their head. There are 4 larval stages separated by a moult (or exuvia) and an adult stage. The larvae closely resemble the adults of the wingless form (Fig. 5). In the winged form, the wings develop progressively from stage to stage. A single colony may contain both winged and wingless forms. Winged forms appear mainly in the event of overpopulation and then migrate to other plants (Leclant, 1999).

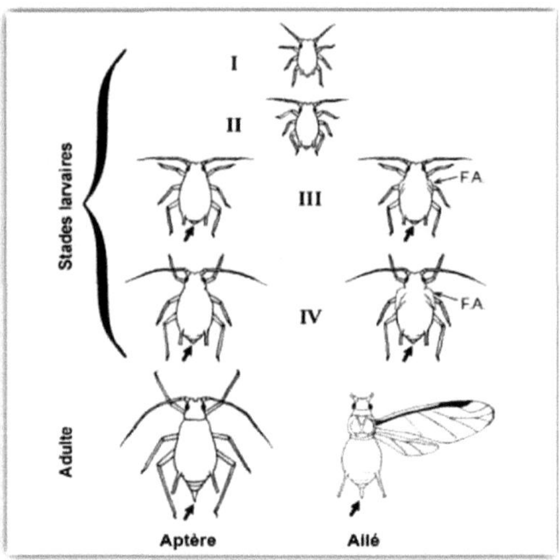

Figure 5: Aphid development stages (Godin and Boivin, 2000)

2.3. Biology

Many aphid species have a primary host (shrubs) and secondary hosts (herbaceous plants). In autumn, after mating, the female lays winter eggs on the primary host and, for the rest of the year, populations are made up entirely of viviparous and parthenogenetic females (giving birth to young larvae without the need for mating). However, under cover, aphids can remain on their secondary hosts all year round. When the species is specific, the females can lay eggs on this host. The population develops in foci either from infested plants or from adults. Primary foci spread from plant to plant, and the winged aphids that appear cause a general spread of aphids across the crop (Simon, 2007).Aphids, especially in the case of Aphis, are often spotted by ants looking for honeydew on plants or by observing moult remains (exuviae) (Robert, 2008). Under shelter conditions, aphids can multiply very quickly. Development time is strongly influenced by temperature, and at 20°C lasts around 1 to 2 weeks (Robert,

2008).Aphids can be found on leaves (upper and lower surfaces), in the heart of plants, on stems, runners, flowers and fruit (Simon, 2007).

2.4. Damage

Aphids feed exclusively on plants. They insert their stylets into the tissues and extract sap after emitting saliva, which is sometimes toxic. They weaken the plant and excrete droplets of honeydew, a sugary substance on which blackish fungi or fumagine can develop, limiting photosynthesis and commercially depreciating the affected fruit (Leclant, 2001). Damage is linked to population density and their location on the plant. During their stings, aphids are capable of transmitting viral particles. They play a major role in the spread of virus diseases. As such, they are dangerous because only a few individuals are needed to cause damage (Simon et al, 2007).

2.5. Pistachio aphids from l'Atlas

Seven aphids specific to the host species, subfamily Pemphiginae, genus Fordini (Blackman and Eastop, 1994), induce galls on Pistacia atlantica Desf. trees. Four of these are considered to be the most common in Palestine (Laine, 1995): Slavum wertheimae, Smynthurodes betae West (Photo 2), Geoica utricularia (Photo 3) and Forda riccobonii (Photo 4).

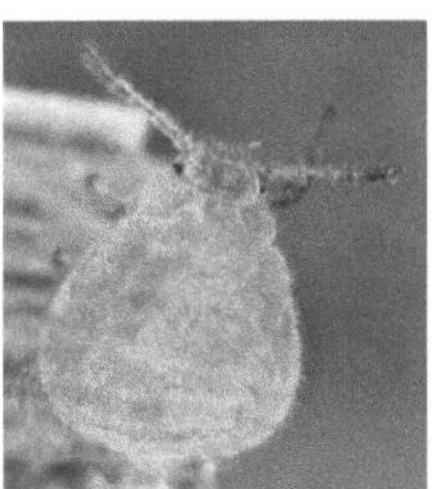

Photo 02: Smynthurodes betae

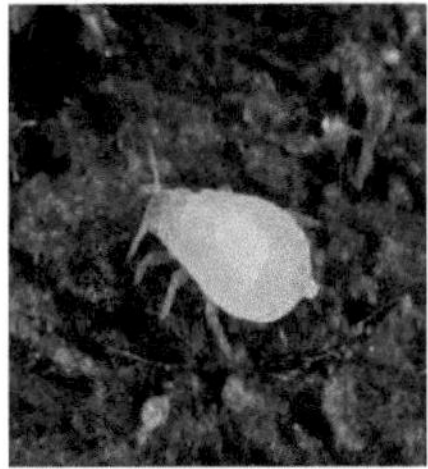

Photo 03: Geoica utricularia

Photo 04: Forda riccobonii

The first two species have a well-known (homocyclic) life cycle (Laine, 1995; 2003).To date, no research has focused specifically on the ecology of the Forda riccobonii population (Martinez, 2009). This aphid has a complex life cycle, forming two different galls on the host tree. spring, is a small red sphere (<5 mm) on the midrib of the tracts (Photo 5), very similar to those formed by Smynthurodes betae.

Photo 05: Small spherical red galls on the leaves of Pistachio trees d'Atlas
Source: original (Mri'gha, February 2017)

The second forms a sequence of a variable number of red spherical chambers articulated on the margins of the leaf (Martinez, 2008). (Photo 6).

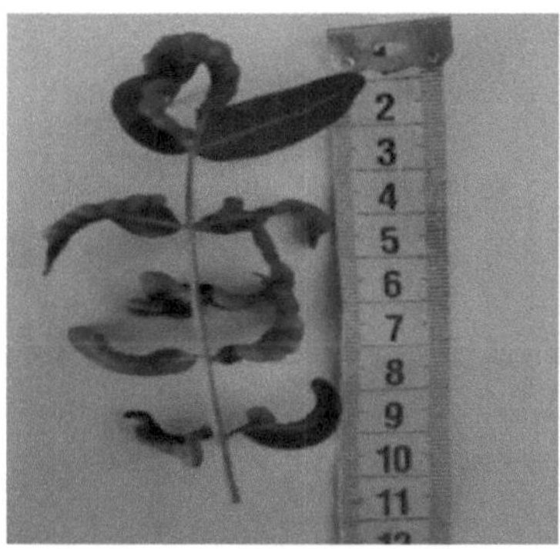

Photo 06: Spherical galls articulated on the leaf margins (Mri'gha, February 2017)

Whether the size of the gall plays a role in the physical shape of the aphids or the rate of attack by enemies. In autumn, the winged aphids of the third generation leave the slit and fly to neighbouring Poaceae (secondary host), where they reproduce partheno-genetically for several generations of Virgogeniae apterae (Wertheim, 1954). The following spring, the winged aphids from the autumn nests fly back to their main host, tracking and producing a sexual generation, which is partnered. The fertilised egg remains in the body of the dead female, persists throughout the year and next spring completes a 2-year cycle (Laine, 2003). Laine (1995) indicated that this aphid species is concentrated in a few trees that it colonises heavily (high density of galls on each colonised tree, which he calls conditional high density).

Martinez and Wool (2003, 2006) showed that trees growing alongside roads were more often parasitized by the Forda riccobonii aphid than trees further away, and that they supported more galls in this disturbed environment.

Martinez (2008) reported that Slavum wertheimae lowers the colonisation rate of other aphid sprays, including Forda riccobonii, by stopping the development of P. atlantica shoots.Inbar and Wool (1995) detected evidence of resource sharing between Forda riccobonii and S. betae. In Palestine, few insect species attack or live inside aphid chicks in Pistacia trees. The atlas pistachio tree, a non-forest tree, is one of these little-known resources. It is only recently that environmental and other services elsewhere in the world are paying more attention to this resource (Bellefontaine, 2001). The Atlas pistachio is one of the most resistant species in arid steppic zones, subject to edapho-climatic constraints on the one hand and parasitic attack on the other.

CHAPTER 3
MATERIALS AND METHODS

The methodological approach used in this study consists of carrying out dendrometric measurements on stands of Atlas pistachio at 3 sites in the Laghouat region. Several outings were carried out at three sites: two stands in gardens in the city of Laghouat and a natural stand (daya) in the El Khneg region. The work consisted in analysing the state of health of the pistachio tree by determining the degree of infestation by aphids, especially the golden aphid.

3.1. Location of the study region :

The wilaya of Laghouat is located 400 km south of Algiers on the Algiers-Ghardaïa road. It lies at an altitude of 767 m on the southern flank of the Saharan Atlas. It covers an area of 25,052 km^2 (Amghar and Kadi Hanifi, 2002) (Fig. 6). It is bordered to the north by the wilaya of Tiaret, to the south by the wilaya of Ghardaïa, to the east by the wilaya of Djelfa and to the west by the wilaya of El Bayadh.

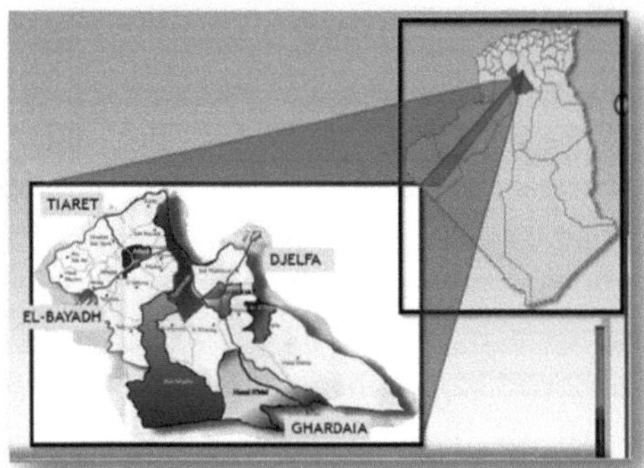

Figure 6: Location of the wilaya of Laghouat

The wilaya is characterised by a relief and formations of high steppe plains covering an area of 1,082,292 ha, i.e. 89% of the wilaya's surface area. The rest of the land is made up of: 100,665 hectares of woodland, with an afforestation rate of 8%; 27,548 hectares of farmland, or 2%, planted with various crops: cereals (barley, durum wheat, common wheat, oats), fodder (alfalfa), vegetables (potatoes, tomatoes, onions) and trees. Finally, unproductive land (dunes, depressions and built-up areas) covers 5,650 ha, or 1% (Salemkour et al, 2013).

The Saharan Atlas zone is characterised by altitudes ranging from 1,000 to over 1,700 m, with slopes ranging from 12.5 to 25%. This area in the north-west of the Wilaya (Aflou and Brida regions) is made up of old forest massifs covering an area of The Hauts Plateaux and Plateaux Sahariens areas are made up of arable land covering 47,095 ha, alfa beds covering 315,125 ha and pastures and rangelands covering 1,531,766 ha. The High Plateaux and Saharan Plateaux zone is characterised by altitudes ranging from 700 to 1,000 m and slopes of 0 to 3% (D.P.A.T, 2010).

3.2. Climate

3.2.1. Rainfall

The highest rainfall was recorded in September (27.02 mm), while the lowest was in July (6.33 mm). The cumulative annual rainfall over 19 years is 165.97 mm (Tab.1).

Table 1: Average monthly precipitation for the period 1996-2014 in the Laghouat region.

Month	J	F	MR	A	M	J	JT	A	S	O	N	D	Avg
P mm)	13,52	6,81	11,49	18,98	10,20	9,65	6,33	12,27	27,02	19,72	19,49	10,49	165,97

Source: ONM, 2015.

In the Laghouat region, annual rainfall is irregular and low, with t h e   lowest value of 62.7 mm recorded in 1998 and the highest value recorded in 2011 at 285.2 mm (Fig. 7).

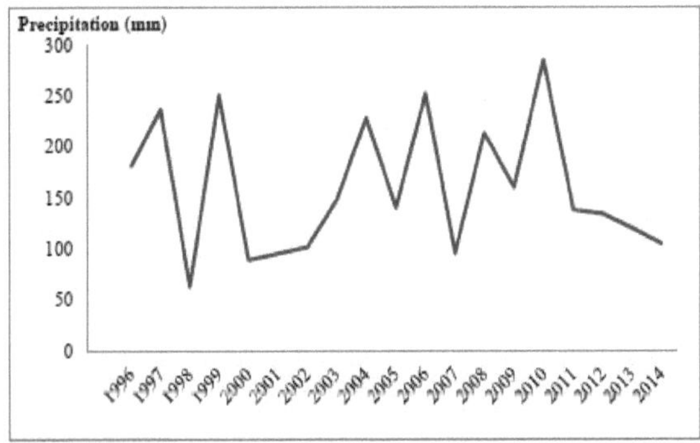

Figure 7: Interannual precipitation in the Laghouat region (1996-2014).

3.2.2. Temperature

The average maximum temperature reached **31.16°C** in July. In January, on the other hand, the average minimum temperature was **8.47°C.** The maximum monthly temperature was 39.68°C in July, while the minimum monthly temperature was 1.84°C in January (Table 2).

Table 2: Average monthly temperature over the period (1996-2014) in the Laghouat region.

Month	J	F	MR	A	M	J	JT	A	S	O	N	D	Avg
T max.(°C)	15,09	16,70	21,03	24,92	29,83	35,70	39,68	39,09	31,12	25,63	18,57	15,31	
T min. (°C)	1,84	2,69	6,01	9,44	14,53	18,99	22,64	22,34	17,75	12,43	5,85	2,55	
Average temperature (°C)	8,47	9,70	13,52	17,18	22,18	27,35	31,16	30,72	24,44	19,03	12,21	8,93	18,74

Source: ONM, 2015.

T max. average maximum monthly temperature in **(°C)**. **T min.** average minimum monthly temperature **(°C)**. **T.avg.**: average monthly temperature in **°C.**

Figure 8 shows that average annual temperatures vary between 17.7°C and 20°C. The coldest year was 2005 at 17.7°C, while the hottest year was 2001 at 20°C.

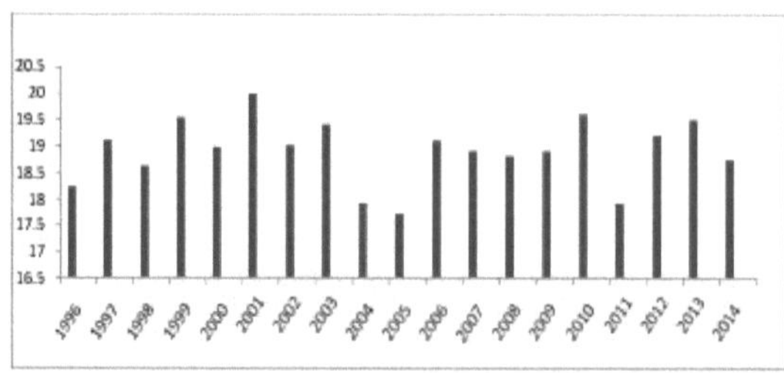

Figure 8: Interannual variation in temperatures in the Laghouat region (1996-2014).

The other climatic parameters provided a clearer picture of the bioclimatic aspect of the region.

3.2.3. Relative humidity of the air

The highest humidity was recorded in December at 49.66%, and the lowest in July at 20.44% (Table 3).

Table 3. Monthly relative humidity for the period 2001-2013.

Month	J.	F.	M.	A.	M.	J.	J.	A.	S.	O.	N.	D.
H %	47,22	41,27	30,61	32,61	28,66	25,33	20,44	22,94	33,33	39,83	45,88	49,66

Source: ONM, 2014

3.2.4. Wind

The frequency and intensity of winds are also a feature of Saharan climatology. They play a considerable role, causing deflation and corrosion on the relief, but they also act on plants, especially on their aerial parts, by accentuating evapotranspiration (Ozenda, 1983). Although the wind itself is nothing exceptional in the desert, its effects are striking. Wind is a major constraint for certain crops in the Sahara. The average wind speed over the period 1996-2013 was 3.43 m/s, with a maximum value of 4.88 m/s in July (Table 4).

Table 4: Monthly wind speeds recorded for the period 1996-2013 in Laghouat

Month	J.	F.	M.	A.	M.	J.	J.	A.	S.	O.	N.	D.	Avg
Wind speed (m/s)	2.96	3.52	3.66	4.37	3.92	3.55	4.88	3.11	3.06	2.49	2.79	2.85	3.43

Source: ONM, 2014

Wind speed is irregular in the Laghouat region, with a maximum value of around 5.6 m/s recorded in 1999 (Figure 9).

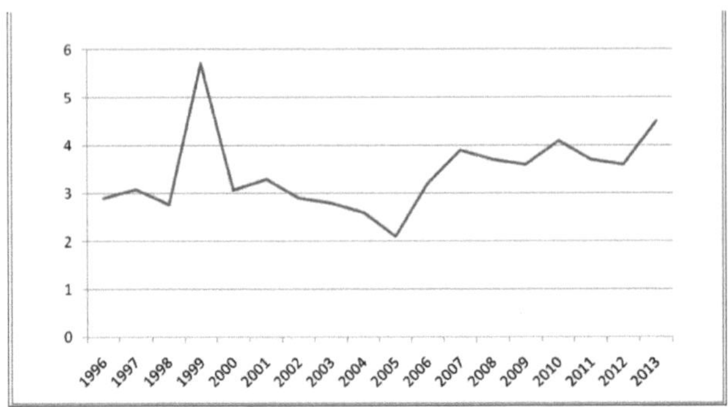

Figure 9: Annual wind speed in the Laghouat region (1996-2013).

a) The sirocco

This is a hot, drying wind, but it does help to increase evapotranspiration. It occurs in the hottest months of the year, and is very common in the summer. This very hot and dry wind blows from the south-west, laden with sand, sometimes lasting for several days and killing plants. the most delicate oases. It also accumulates sand around certain plants and buries them. If they are not cleared, they succumb (Dubief, 1951).

b) The storms of sands

The frequency of sandstorms is directly related to the strength of the winds that blow over the steppic zone during the year. According to Dubief (1959), a sandstorm is a turbulent wind of any strength blowing over an area of at least a few square kilometres, carrying sand particles in such numbers and of such diameter as to obstruct the view of a standing observer wearing sand goggles. This phenomenon is very frequent during the hottest season, particularly in July and August.

3.2.5. Jelly

Frost is a form of atmospheric precipitation that appears as soon as the temperature falls below zero (Bahri, 2007). The table below shows that the frost was very severe in January, lasting 8.25 days (Table 5).

Table 5: Number of days with frost in the Laghouat region (2001-2012).

Month	J.	F.	M.	A.	M.	J.	J.	A.	S.	O.	N.	D.
Number of days with frost	8.25	4.25	0.25	00	00	00	00	00	00	00	0.08	5

Source: ONM, 2013.

3.2.6. Climate summary :

In general, climatic factors do not act in isolation from one another. The bioclimatic stage of a region and its drought period can only be determined from the synthesis of two climatic parameters, such as temperature and rainfall.

6.2.6.1. Umbrothermal diagram :

The Bagnouls and Gaussen umbrothermal diagram enables us to determine the length of the dry period over the course of a year. This dry period is represented by the intersection of the two temperature and rainfall curves. The The diagram is based on the scale: P (mm) = 2T (°C), where P represents monthly rainfall and T the average monthly temperature over the period in question. The umbrothermal diagram shown in the figure above shows that our study area is characterised by a dry period throughout the year (Fig. 10).

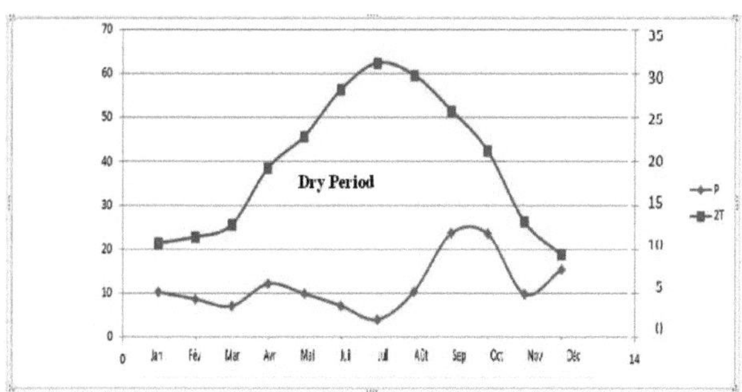

Figure 10: Umbrothermal diagram by BAGNOULS and GAUSSEN (1996-2014)

6.2.6.2. Rainfall quotient and Emberger climagram

Emberger's climagram (1955) can be used to determine the bioclimatic stage of a given station by calculating the rainfall-thermal coefficient using the following formula:

$$Q_2 = 2000\ P/\ (M^2 - m)^2$$

This formula was simplified by **STEWART in 1969**.

$$Q_3 = 3.43\ P/\ (M\text{-}m)$$

where :

Q_3 :   rainfall quotient.

P: average annual precipitation (mm).

M: maximum mean monthly temperatures (°C).

m: minimum average monthly temperatures (°C).

The result obtained (Q_3 = 15.04) with m=1.84°C means that Laghouat is classified as having a Saharan bioclimate, with a cool winter variant (Fig. 11).

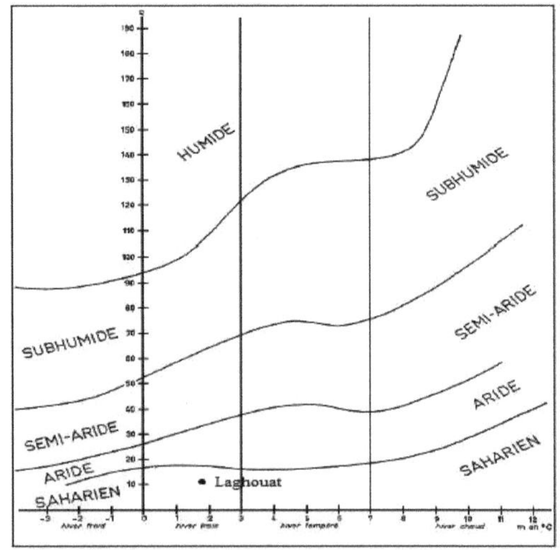

Figure 11: Emberger climagram for the Laghouat region

6.2.6.3. De Martonne aridity index

This index takes into account annual rainfall and temperature. De Martonne's annual aridity index is explained by the formula: P/(T+ 10) where P is the annual rainfall (in mm) and T is the mean annual temperature (in °C) (Table 6).

Table 06: Aridity index (I) values and corresponding bioclimates.

Valeur de l'indice	Type de bioclimat
0<I<5	Hyperaride (HA)
5<I<10	Aride (A)
10<I<20	Semi-aride (SA)
20<I<30	Subhumide (SH)
30<I<55	Humide (H)
I>55	Perhumide (PH)

Calculating this index for the Laghouat region classifies it as an arid climate with I=5.77.

3.3. Geographical location of stations

The three study stations are located in or near Laghouat (Fig. 12).

Caption:

Station 1: Amusement and Leisure Park (M'righa) Station 2: Snober Garden
Station 3: Daya de Kheneg

Figure 12: Location of sampling stations (Google maps, 2017)

The first station is the Jardin Public de Snober with coordinates: 33.809108 N; 2.871023 E and 720 m altitude. It is located within the city of Laghouat and covers an area of 07 ha. This public garden has a children's play area and family areas. The dominant species are Aleppo pine, casuarina, olive and more than fifty Atlas pistachio trees. These trees are regularly irrigated by the agents of the APC of the wilaya of Laghouat (photo 7). However, the second station is that of the former Mrigha amusement park with coordinates: 33.809108 N; 2.871023 E and 750 m altitude. It is located on the northern outskirts of the town of Laghouat, and is an amusement and leisure park covering an area of 87 hectares, 40 hectares of which are wooded with mostly exotic forest species, such as Eucalyptus (several species including E. globulus, E. camaldulensis....), the false pepper tree: Schinus molle, Melia azedarach, Biota orientalis, Cupressus sempervirens, Acacia retinoïdes, Acacia horrida, Acacia cyanophylla and Pistacia atlantica... etc (photo 7). The third station is a daya, halfway between the commune of Kheneg and El Houita, known as Dayat nosse, is 33.809108N, 2.871023E and 780m above sea level. It is located 7 km south of the town of Laghouat. The daya is of the slightly depressed type (Pouget, 1980) (photo 7).

Site 1: Snobar GardenSite 2: Mrigha Garden Site 3: Daya Kheneg

Photo 7: Panoramic view of the study sites

Our choice of three sites aims to reveal the differences, if any, between the effects of aphids on these different natural and artificial stands.

3.4 Methodology

In this study, we characterised the three Pistachio stands at the three sites from a biometric point of view in an attempt to establish the link between these parameters and aphid damage.

3.4.1. Dendrometric measurements :

The main dendrometric parameters taken into account are :

a) Tree height

We estimated the height using a simple stick with a fairly high degree of accuracy (Massenet, 2005) (Fig. 13).

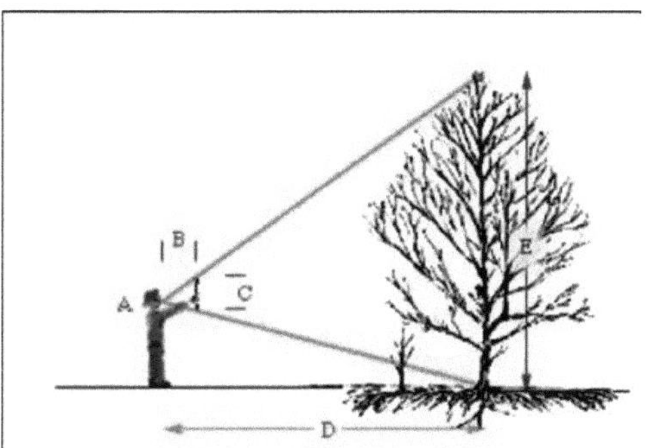

Figure 13: Simple estimation of tree height

A - Observer

B - Distance between the observer's eye and the measuring stick

C - Height of the measuring stick

D - Distance from the observer to the tree

E - Height of tree

The height (E) of the shaft is equal to the distance to the shaft (D) multiplied by the height of the stick (C) divided by the length of the arm (B). E=D*C/B

b) Circumference at 1.3 m

Circumference is generally measured using a tape, which should be non-deformable if possible, with a metal weave or, better still, a fibreglass weave. This 1.5 m or 3 m tape can be used to measure at all levels: the measurement at chest height is considered to be 1.3 m (Massenet, 2005).

c) Number of first branches

We counted the number of first branches on the tree trunk, which gives an indication of the tree's vigour and health.

d) Height of trunk

We measured the height of the trunk (from ground level to the first branch) using a tape measure.

e) Measures relating to the crown

Vigour indices seek to quantify a tree's vitality and social status using, for example, some of its aerial dimensions (IPGRI, 1997). Crown indices use information on the characteristics of tree crowns (volume, surface area, etc). Measurements of the width and length of the crown were taken in order to calculate its volume.

e.1) Tree crown volume :

After estimating the length and width of the crown, we calculated the volume of each tree using the following formula:

$$\mathbf{Volume_{arbre}} = 1\ /2(4\ /3\ \pi\ \mathbf{R^3} * \pi * \mathbf{R^3}\quad \mathbf{R} = \text{Height of the crown}; \pi = 3.14$$

f) Descriptors for leaves

The size of the leaves is characteristic of the species or cultivar in question, but c a n vary considerably depending on the vigour of the shoot (IPGRI, 1997).
For the following descriptors, an average of 10 fully developed representative leaves collected from different trees (IPGRI, 1997).

f.1. Leaf length [m]

Measured from the base of the petiole to the top of the terminal leaflet (Fig. 14).

f.2. Leaf width [m]

Measured at its widest point (Fig. 14).

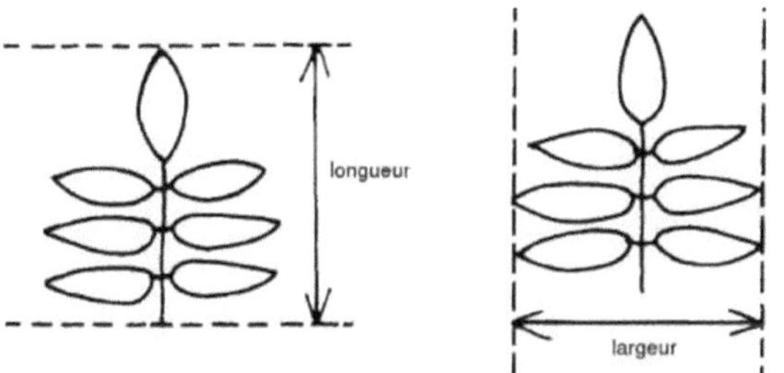

Figure 14: Length and width of Atlas pistachio leaves

f.3. Length and width of leaflet terminal

The length and width of the terminal leaflet are measured at its widest point (IPGRI, 1997) (Fig. 15).

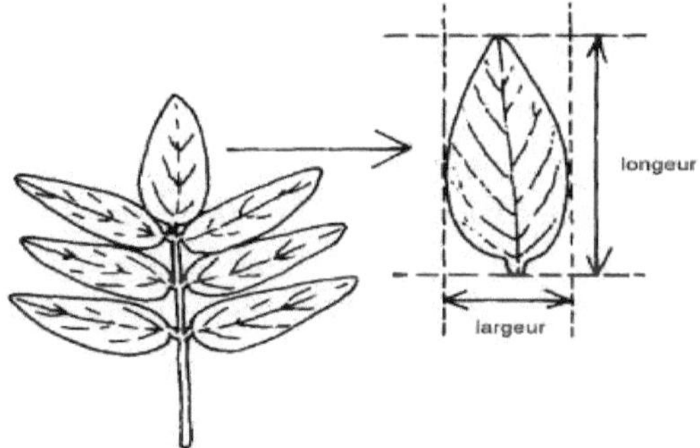

Figure 15: Length and width of the terminal leaflet

3.4.2. Sex ratio :

The three sites studied show similar rates between males and females (Tab 7.).

Table 7: Sex distribution of male and female plants at the three sites

Gender	Mrigha	Snober	Kheneg
Male	22,22	29,63	23,08
Female	77,78	70,37	76,92

3.4.3. Number of galls per tree :

A gall is a tumorous growth produced on the stems, leaves, buds or fruit of certain plants as a result of the stings of parasitic animals (Dauphin, 1993). We counted the number of galls on the foliage of each pistachio tree in a randomly selected area of 0.125 m^3 (0.5 * 0.5 * 0.5 m). The total number of galls was then estimated for each tree in relation to its crown volume.

3.4.4. Analysis of galls

Fresh samples are transported to the laboratory and stored in the refrigerator for a short period, after which the galls are dissected. Once these galls were opened using dissection equipment, we preserved the aphids found in 2 ml Eppendorf tubes with 75% Ethanol.

3.4.4.1. Identification

a. Macroscopic description

Macroscopic observation is a technique used to describe the morphology, colour, contour, surface appearance and size of galls.

b. Microscopic observation

Microscopic observation of aphids involves studying the morphology of the aphid, including the body, the pair of cornicles, the more or less long cauda (tail at the end of the abdomen), the antennae, the frontal sinus and the rostrum (sucking stylet). The presence or absence of wings can also be observed.

Our samples were identified under the supervision of Mr Laamari Malik from the Hadj Lakhdar University in Batna and Mr Nicolas Perez Hidalgo from the University of Valencia in Spain.

3.4.5. Statistical analysis

Statistical data processing involved descriptive statistics for the various measurements, analyses of variance to compare the three sites and correlations between the parameters studied. We used Statistix 8 under Windows for the statistical analyses.

PART II

RESULTS AND DISCUSSION

4.1. Tree dendrometric parameters

4.1.1 Tree height :

A total of 79 trees were sampled on the three sites. The average height of these trees was 7.88 m ± 0.4 m, with a minimum of 1.6 m and a maximum of 17 m, The average height measured in the M'righa park is 2.79 m ± 1.64 and varies between 1.6 m and 9 m, while the average height of the trees in the Kheneg park is 10.75 m ± 0.27 and varies between 6.8 and 16 m, and for the Snober garden the average height of the trees is 10.1 ± 0.31 and varies between 5 m and 17 m (Fig. 16).

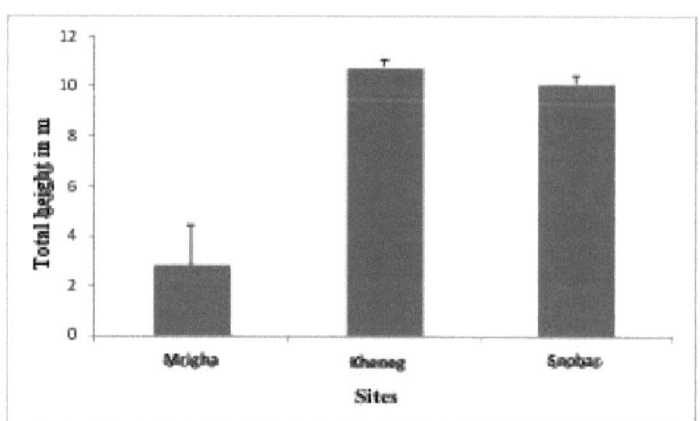

Figure 16: Average tree heights measured at the three sites Analysis of variance on tree height values at the three sites indicates a significant difference (F $_{,278}$ = 209; P ≤ 0.00001); the trees at M'righa h a v e lower heights than those at Kheneg and Snober.

The main significant correlations recorded are s h o w n in table (8).

Table 8: Main significant correlations recorded with total tree height

Parameter Correlation

Trunk height.81; p≤0.00001; N = 76

Leaf widthr=0.77; p≤0.00001; N = 76
Width of terminal leafletsr=0.74; p≤0.00001; N = 76

Tree crown width 76 ; p≤0.00001; N = 76

Leaf lengthr=0.62 ; p≤0.00001; N = 76
Length of terminal leaflets

Number of first branches r=0.48; p≤0.00001; N = 76 r=0.40; p=0.0002; N = 76

Tree crown volume.35 ; p=0.0012; N =76

Number of gallsr=0.40 ; p=0.0002; N = 76

4-1-2 Circumference at 1.3 m :

We found a significant difference between the circumferences at 1.3 m of the trees in the three sites (F2.78= 1.38; P=0.2581). The average circumference at 1.3 m measured in the three sites is 0.26 m ± 0.015 and varies between 0.04 m and 2.4 m.The average circumference at 1.30 m measured in the M'righa park is 0.11 m. ± 0.01 and varies between 0.040 and 0.58 m, the average circumference at 1.3 m of the trees in the Kheneg daya is 1.4 m ± 0.027 m and varies between 0.70 m and 2.9 m. The average circumference of the trees in the Snober garden is 1.12 m ± 0.047 m and varies between 0.2 m and 2.4 m (Fig. 17).

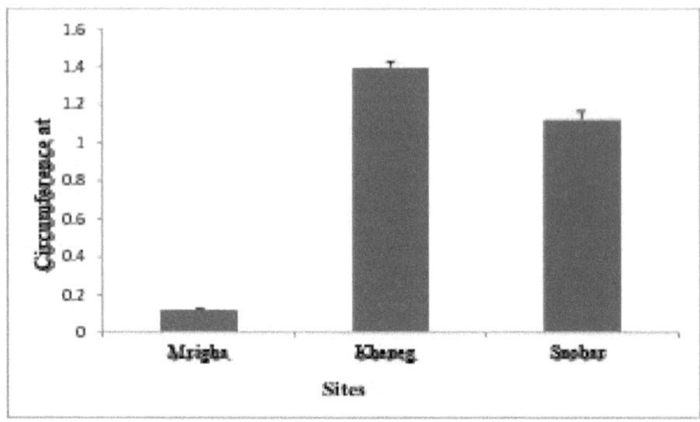

Figure 17: Circumference at 1.30 m of trees measured at the three sites

The trees at M'righa have smaller circumferences than those at Jardin Snober and Kheneg (Fig. 17). There was a positive and statistically significant correlation between the circumference of the tree at 1.3 m and the number of first branches (r=0.49; P ≤ 0.00001).

4-1-3 Number of first branches :

The average number of first branches counted in the three sites is 2.97 branches ± 0.17 and varies between 1 and 10 branches. The average number of first branches counted in the M'righa park is 2.34 branches ± 0.141 and varies between 1 and 6 branches per stand. However, the average number of first branches in the El Kheneg daya is 3.50 branches ± 0.2 and varies between 2 and 10 branches. Whereas the number of branches in the trees in the Snober garden is 3.07 branches ± 0.141 and varies between 1 and 6 branches (Fig. 18).

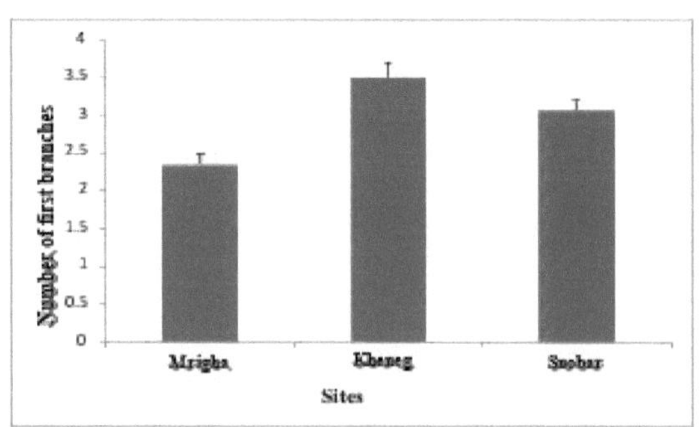

Figure 18: Number of first branches at the three sites

We found a statistically significant difference between the number of first branches in the three sites (F2, 78= 3.23; P = 0.045).

The main significant correlations recorded are s h o w n in table (9).
Table 9: Main significant correlations recorded with the number of first branches.

Parameter Correlation

Trunk height.4; p=0.0002; N = 76

Tree width.34; p=0.0002; N = 76

Leaf width.31; p=0.0005; N = 76

Tree height .40; p=0.0002; N = 76

Number of galls.32; p=0.0034; N = 76

Circumference at 1.3 mr=0.49; P ≤ 0.00001; N = 76

4-1-4 Trunk heights :

The average height measured at the three sites was 2.98 m ± 0.18 m, varying between 0.4 and 7 m. The average height of the trunk measured in the M'righa park is 0.97 m ± 0.04 and varies between 0.4 and 2 m. It is 3.94 m ± 0.14 m in the Kheneg daya and varies between 1.5 and 6.5 m. However, the average height of the trunks measured in the Snober garden is 04 m ± 0.12 and varies between 2 and 7 m. We found a significant difference between the trunk heights at the three sites (F2.78= 62; P ≤ 0.00001), with the trees at M'righa showing lower heights than those at Jardin Snober and Kheneg (Fig. 19).

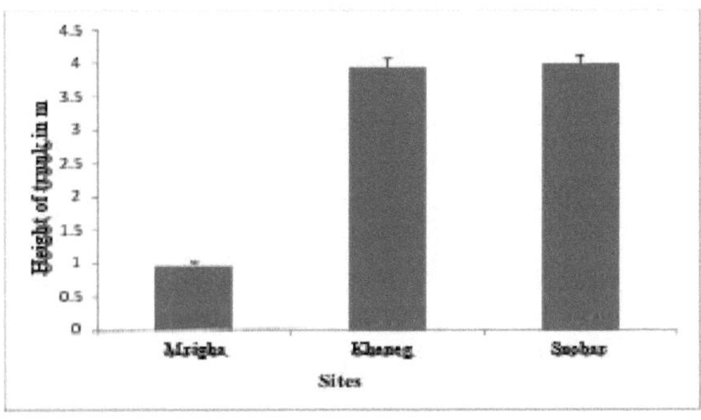

Figure 19: Trunk height measured at the three sites

The main significant correlations recorded are shown in table (10).

Table10: Main significant correlations recorded with total trunk height

Parameter	Correlation
Leaf width	r=0.70; P ≤ 0.00001; N = 76
Width of terminal leaflets	r=0.80; P ≤ 0.00001; N = 76
Tree crown width	r=0.75; P ≤ 0.00001; N = 76
Leaf length	r=0.50; p≤0.00001; N = 76
Length of terminal leaflets	r=0,42 ; p=0, 00001; N = 76
Number of galls	r=0.34; p=0, 0016; N = 76

4-1-5 Height of crown :

The average height measured in the three sites is 4.8 m ± 0.03 and varies between 0.6 and 14.5 m. The average crown height measured in the M'righa park is 1.82 m ± 0.10 and varies between 0.6 and 7 m, and in the Daya Kheneg is 6.81 m ± 0.14 and varies between 2.5 and 10.5 m. However, the average crown height measured in the Snober garden is 06 m ± 0.28 and varies between 1 and 14 m (Fig. 20).

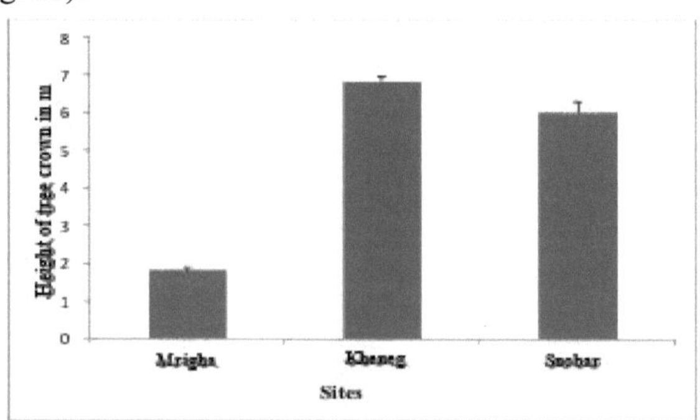

Figure 20: Tree crown height measured at the three sites

We found a highly significant difference between crown heights at the three sites ($F_{2,78}$ = 37.9; P ≤ 0.00001), with trees at Kheneg and Snober having higher crowns than those at M'righa .The main significant correlations recorded are shown in table (11).

Table 11: Main significant correlations recorded with crown height

Parameter	Correlation
Height of tree	r=0.94; p≤0.00001; N = 76
Height of trunk	r=0.58; p≤0.00001; N = 76
Leaf width	r=0.69; p≤0.00001; N = 76
Width of trees	r=0.64; p≤0.00001; N = 76
Leaf length	r=0.56; p≤0.00001; N = 76
Length of terminal leaflets	r=0.43; p=0.0001; N = 76
Tree crown volume	r=0.35; p=0.0012; N = 76
Number of galls	r=0.37; p=0.0007; N = 76

4-1-6 Tree crown volume :

The average crown volume calculated at the three sites is 2546.6 m^3 ± 0.4 and varies between 0.6 and 28649 m3.The average crown volume calculated in the M'righa park is 12.65 m3 ± 0.37 and varies between 0.6 and 18.8 m3 , and in the Kheneg daya is 2270.2 m3 ± 22 and varies between 55.95 and 12086 m^3 . However, the average crown volume measured in the Snober garden is 5252.5 m^3 ± 0.02 m^3 and varies between 88.55 and 28649 m^3 (Fig. 21).

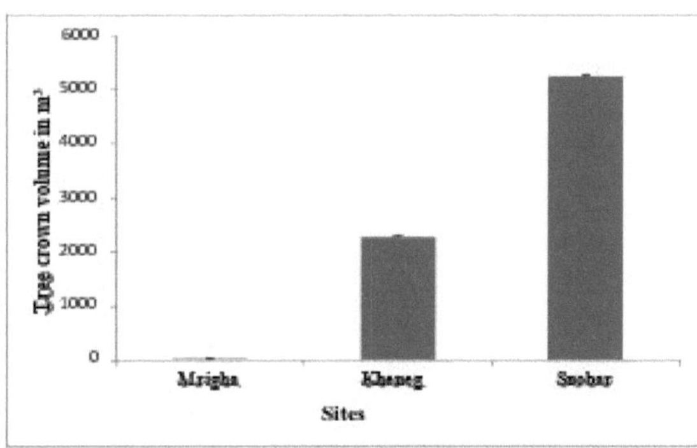

Figure 21: Tree crown volume measured at the three sites

We found a significant difference between crown volumes at the three sites (F $_{2,78}$ = 10.4; p≤0.00001), with the trees at M'righa having smaller crowns than those at Jardin Snober and Kheneg.

The main significant correlations recorded are shown in table (12).

Table 12: Main significant correlations recorded with crown volume

Parameter	Correlation
Height of trunk	r=0.49; p≤0.00001; N = 76
Height of tree	r=0.45; p≤0.00001; N = 76
Leaf width	r=0.50; p≤0.00001; N = 76
Tree crown width	r=0.84; p≤0.00001; N = 76
Length of terminal leaflets	r=0,37 ; p=0, 00007; N = 76
Leaf length	r=0.40; p=0.0002; N = 76
Tree crown height	r=0.35; p=0.0012; N = 76

4-2 Leaf biometry

4-2-1. Leaf length :

The average length of the leaves measured at the three sites is 5 cm ± 0.03 and varies between 3 and 26 cm. The average length of the leaves measured in the M'righa park is 6 cm ± 0.02 and varies between 3 and 13 cm, and in the El Kheneg daya, the average is 10 cm ± 0.02 and varies between 6 and 16 cm. However, the average leaf length measured in the Snober garden was 13 cm ± 0.04 and varied between 9 and 26 cm (Fig. 22).We found a significant difference between leaf lengths at the three sites ($F_{2,78}$ = 30.5; P ≤ 0.00001), with trees at M'righa having lower heights than those at Jardin Kheneg and Snober.

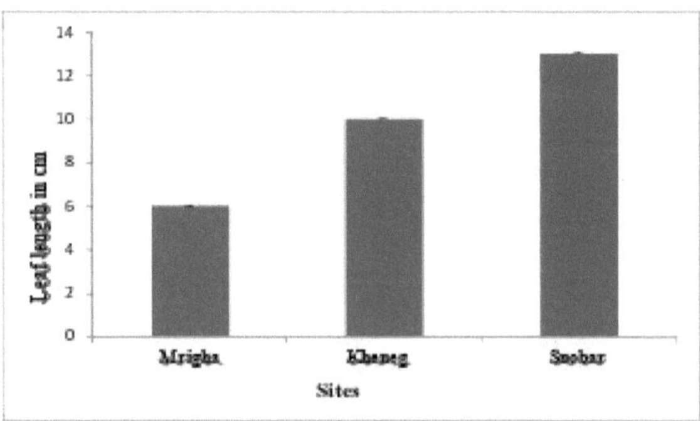

Figure 22: Leaf lengths measured at the three sites

The main significant correlations recorded are shown in table (13).

Table 13: Main significant correlations recorded with leaf length e

Parameter	Correlation
Height of crown	r=0.56; p≤0.00001; N = 76
Height of tree	r=0.62; p≤0.00001; N = 76
Height of trunk	r=0.55; p≤0.00001; N = 76
Leaf width	r=0.75; p≤0.00001; N = 76
Tree crown volume	r=0.40; p = 0.002; N = 76

4-2-2. Sheet width :

The average width of the leaves measured at the three sites was 5 cm ± 0.03, varying between 3 cm and 15 cm. The average width of the leaves measured in the M'righa park is 4 cm ± 0.08 and varies between 3 and 8 cm, and in the El Kheneg daya is 7 cm ± 0.02 and varies between 3 and 15 cm. The average width of the leaves measured in the Snober garden was 8cm ± 0.01 and varied between 5 and 12cm (Fig. 23).

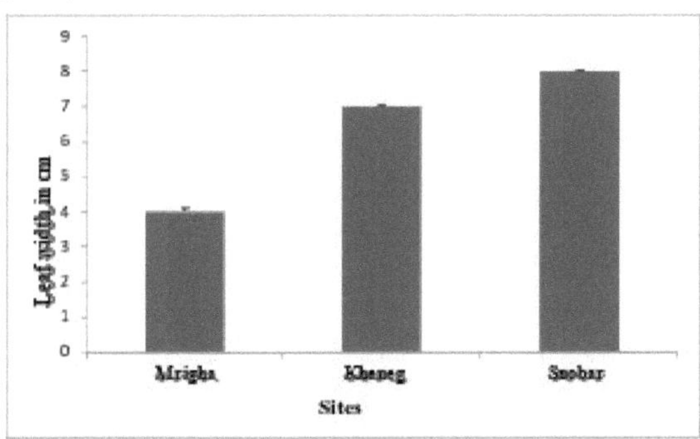

Figure 23: Leaf widths measured at the three sites

We found a highly significant difference in leaf width between the three sites (F2,78= 139; P ≤ 0.00001), with trees at M'righa having lower heights than those at Jardin Snober and Kheneg.

The main significant correlations recorded are shown in table (14).

Table 14: Main significant correlations recorded with leaf width

Parameter Correlation

Width of terminal leaflets r=0.82; p≤0.00001; N = 76

Tree crown width.77 ;p≤0.00001; N = 76

Leaf lengthr=0.75 ; p≤0.00001; N = 76
Length of terminal leaflets

Number of first branchesr=0.31; p≤0.00001; N = 76 r=0.31; p=0.0054; N = 76

Tree crown volumer=0.50; P ≤ 0.00001; N = 76

4-2-3. Width of terminal leaflets :

The average width of the terminal leaflets measured at the three sites is 0.7 cm ± 0.03 and varies between 0.3 cm and 1.3 cm. The average width of the terminal leaflets measured in the M'righa park is 0.65 cm ± 0.02 and varies between 0.3 and 1.3 cm, and in the El Kheneg daya is 0.9 cm ± 0.01 and varies between 0.3 and 1 cm. However, the average width of the terminal leaflets measured in the Snober garden was 0.9 cm ± 0.03 and varied between 0.4 and 1 cm (Fig. 24).

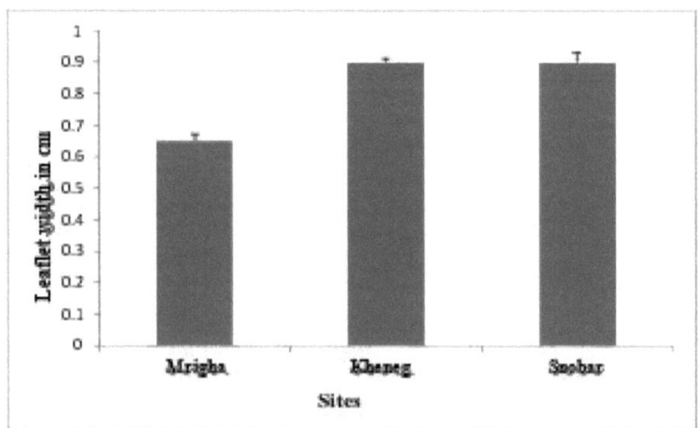

Figure 24: Widths of terminal leaflets measured at the three sites There was a significant difference between the widths of the terminal leaflets and the widths of the terminal leaflets. terminal leaflets in the three sites ($F_{2,78} = 195$; P ≤ 0.00001), the trees in M'righa have smaller terminal leaflet widths than those in the Snober and Kheneg gardens.

The main significant correlations recorded are shown in table (15).

Table 15: Main significant correlations recorded with the width of terminal leaflets

Parameter	Correlation
Height of tree	r=0.74; p≤0.00001; N = 76
Height of trunk	r=0.80; p≤0.00001; N = 76
Height of crown	r=0.59; p≤0.00001; N = 76
Leaf width	r=0.82; p≤0.00001; N = 76
Tree crown width	r=0.77; p≤0.00001; N = 76
Leaf length	r=0.75; p≤0.00001; N = 76
Tree crown volumes	r=0.50; p=0.0002; N = 76

I-2-4. Length of terminal leaflet :

The average length of the terminal leaflet measured at the three sites was 1.09 cm ± 0.01 and varied between 1.5 and 6 cm. The average length of the terminal leaflets measured in the M'righa park is 2 .02 cm ± 0.08 and varies between 1.5 and 4.8 cm, and in the Kheneg daya is 3.26 cm ± 0.09 cm and varies between 1.5 and 4.8 cm. However, the average length of the terminal leaflets measured in the Snober garden was 4.54 cm ± 0.08 cm and ranged from 3.4 to 6 cm (Fig. 25).

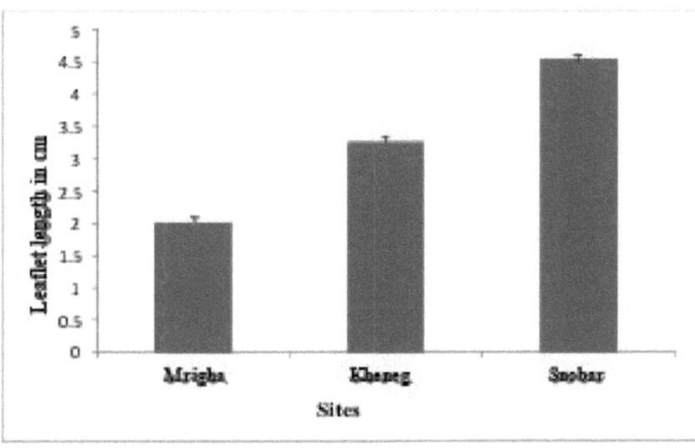

Figure 25: Lengths of terminal leaflets measured at the three sites There was a significant difference between the lengths of the terminal leaflet and the terminal leaflet.terminal in the three sites (F2,78= 315; P ≤ 0.00001), the trees at M'righa have lower heights than those at Kheneg and Snober .

The main significant correlations recorded are shown in table (16).

Table 16: Main significant correlations recorded with terminal leaflet length

Parameter	Correlation
Height of tree	r=0.48; p≤0.00001; N = 76
Height of trunk	r=0.42; p≤0.00001; N = 76
Height of crown	r=0.43; p≤0.00001; N = 76
Leaf width	r=0.31; p≤0.00001; N = 76
Width of terminal leaflets	r=0.31; p=0.0042; N = 76
Number of galls	r=0,42; p=0, 0001; N = 76

4-2-5. Sex ratio and infestation rate

Monitoring of both sexes of pistachio tree revealed that female plants were the most affected by the golden aphid (Fig.26).

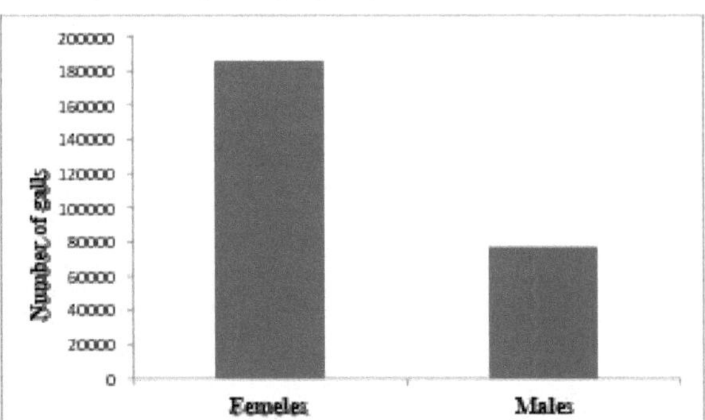

Figure 26: Number of infestations in both sexes

4-2-6. Number of galls :

The average number of galls per plant estimated in the M'righa park is 2.53 galls per plant ± 0.08 and varies between 0 and 36 galls per plant. However, in the Daya Kheneg it is 16.6 galls per plant ± 0.25 and varies between 0 and 80 galls per plant. However, the number of galls per tree in the Snober garden was 4.11 ± 0.14 galls per tree, ranging from 0 to 36 galls per tree (Fig. 27).

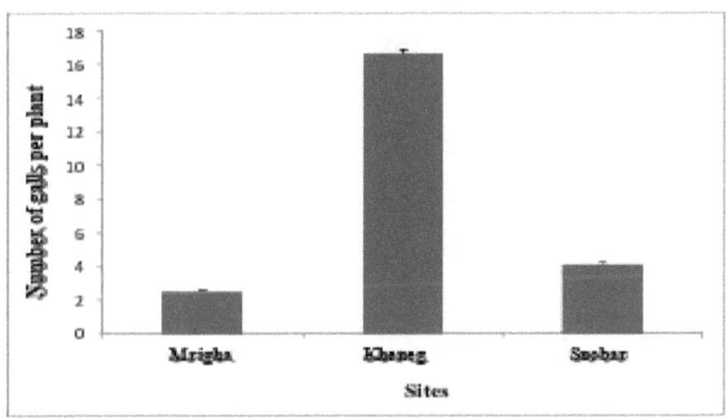

Figure 27: Average number of galls per plant estimated at the three sites

The average number of galls estimated in the three sites is 165291± 0.4 and varies between 0 and 3384674 galls. The average number of galls/site estimated in the M'righa park is 2346.3 galls/site ± 10.67 and varies between 0 and 54388 galls, and in the Kheneg daya is 419511 galls/site ± 83.96 and varies between 0 and 3384674 galls. However, the number of galls per tree site in the Snober garden is 77395 ± 0.141 galls per site, varying between 0 and 773519 galls (Fig. 28).

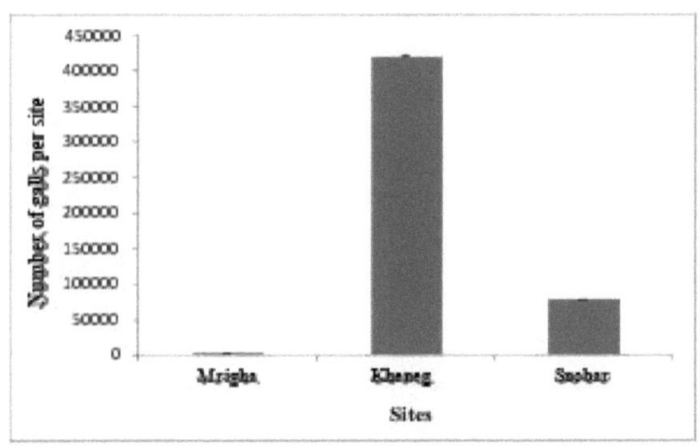

Figure 28: Average number of galls calculated per site

We found a significant difference between the number of galls at the three sites ($F_{2,78}$ = 5.31; P = 0.0047); the trees at Khneg had higher numbers of galls than those at Jardin Snober and M'righa.

The main significant correlations recorded are shown in table (17).

Table17: Main significant correlations recorded with the number of galls

Parameter	Correlation
Height of tree	r=0.40; p=0.0002; N = 76
Height of trunk	r=0.34; p=0, 0016; N = 76
Height of crown	r=0.37; p= 0.0007; N = 76
Leaf width	r=0.34; p=0.00019; N = 76
Length of terminal leaflets	r=0.42; p≤0, 00001; N = 76

4-2-7 Infestation rate :

The three sites studied showed varying degrees of infestation of Pistachio by aphids (Tab.18).

Table 18: Degree of infestation of Pistachio by aphids at the three sites

Gender	No. of feet attacked	Total number of feet	Infestation rate (%)
M'righa	4	26	15,38
Snober	8	28	28,57
Kheneg	14	25	56
Total	26	79	32 ,91

The Kheneg site is heavily infested, with 14 infected trees out of 25, giving a rate of 56%. The Snober public garden is moderately infested, with 8 infected trees out of 28, at a rate of 28.57%, while the young pistachio stand in the M'righa amusement park is weakly infested, with only 4 infected trees out of 26, at a rate of 15.38%.

4-3. Relationship between biometric parameters and galls

4-3-1 Relationship between tree height and number of galls :

The relationship between tree height and the number of galls was assessed (Fig. 28). Pistachio trees of greater height are the most affected by aphids (Fig. 29).

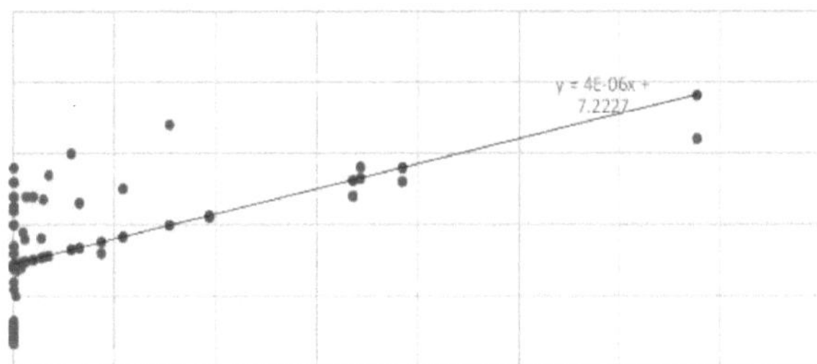

Figure 29: Relationship between tree height and number of galls

4-3-2. Relationship between trunk height and number of galls :

The relationship between trunk height and the number of galls is statistically significant, with galls being more common in trees with long trunks (Fig. 30).

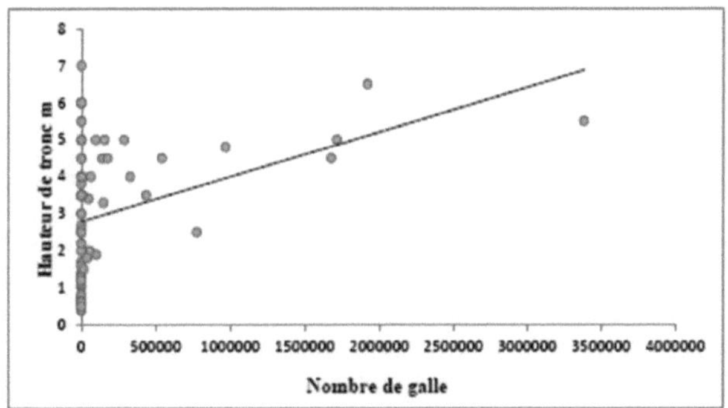

Figure 30: Relationship between trunk height and number of galls

4-3-3. Relationship between crown height and number of galls :

There is a relationship between the height of the crown and the number of galls. Pistachio trees with high crowns are the most affected by aphids (Fig.31).

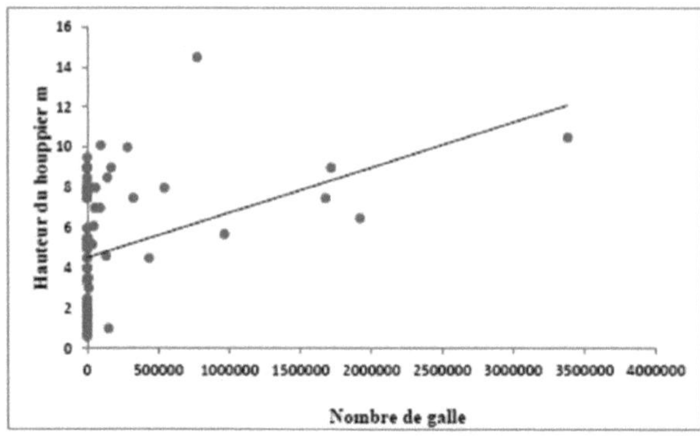

Figure 31: Relationship between crown height and number of galls

4-3-4Relation between leaf width and number of galls :

The relationship between leaf width and the number of galls has been proven. The widest leaves are the most affected by aphids (Fig.32).

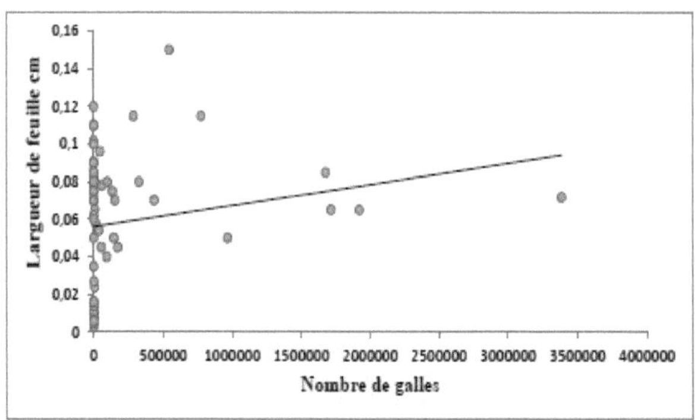

Figure 32: Relationship between leaf width and number of galls

4-3-5. Relationship between the length of terminal leaflets and the number of galls :

There is a relationship between the length of the terminal leaflets and the number of galls; the latter are more numerous in leaves with longer terminal leaflets (Fig.33).

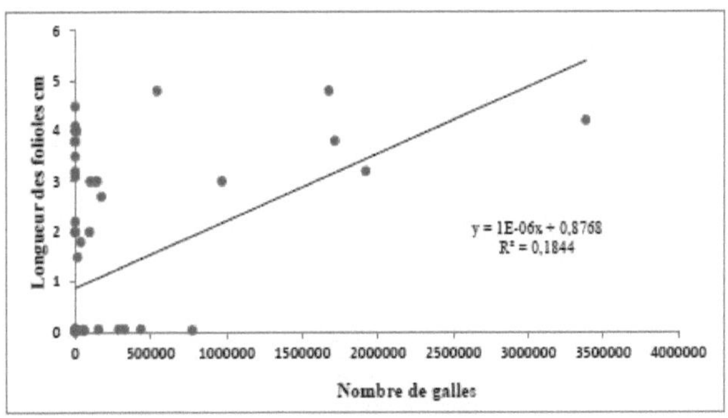

Figure 33: Relationship between the length of terminal leaflets and the number of galls

4-4 - Gall shapes :

We noted the presence of several types of galls, red and green in colour and round or articulated in shape, around 1cm in diameter and sometimes more yellow in colour (photos 8 to 10). Golden aphid galls are very elongated (up to 5 cm or more) and 1.5 to 2 mm in diameter.

Photo 08: Healthy pistachio leaf

Photo 09: Completely deformed leaf

Photo 10: Red spherical galls on pistachio leaves

Small spherical and articulated red galls were observed on the leaves of pistachio trees in the M'righa park on female plants with a very heavy attack compared to male plants. This form of gall was observed only at the M'righa site (Photo 11).

Photo 11: Red articulated spherical galls on pistachio leaves

Source: Originale (Mrigha, February 2017)

The spherical and green articulated galls on the pistachio leaf in the Mrigha garden and the Kheneg daya are found on the male plants, which are more affected by the green articulated galls than the female plants. We noticed that the Kheneg daya was the most affected by the aphid compared with the other two sites (photos 12; 13).

Photo 12: Green spherical galls on pistachio leaves

Source: Originale (Kheneg, February 2017)

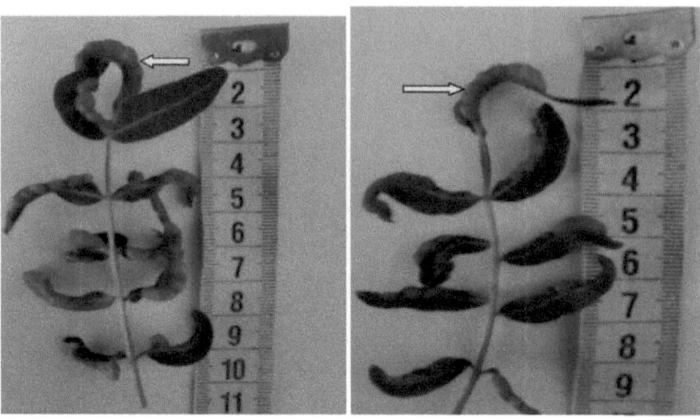

Source: Originale (Kheneg, February 2017)

The different shapes, positions and colours of galls encountered at the three sites enabled us to confirm the presence of the golden aphid (Forda riccobonii) and to identify more other aphid species (photos 14, 15 and 16).

Photo 13: Forda riccobonii gall utricularia on Atlas pistachio leaves

Photo 14: Gall of Geoica on Atlas pistachio leaves

Photo 16: Gall of Smynthurodes betae on Atlas pistachio leaves

Observations in the laboratory revealed several stages of golden aphid (photos 17 to 19).

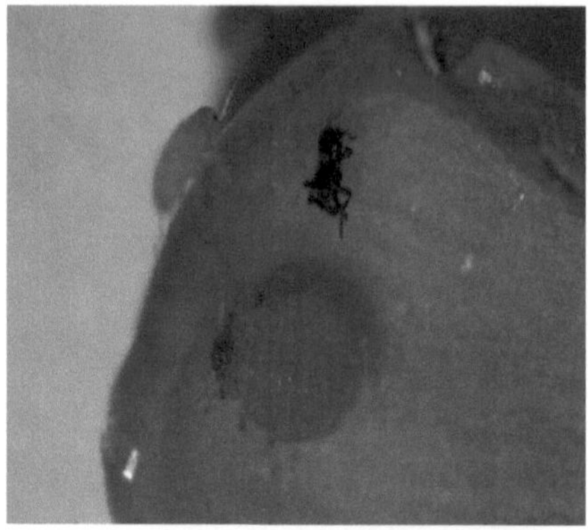

Photo 17: Observation of an aphid golden dorsal side inside a gall

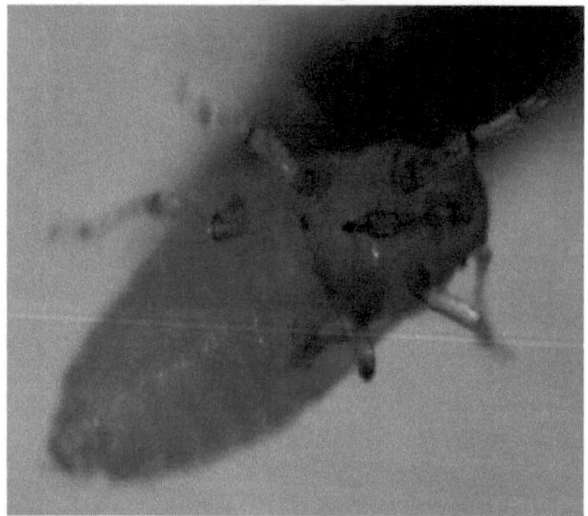

Photo 18: Observation of an aphid ventral side inside agall

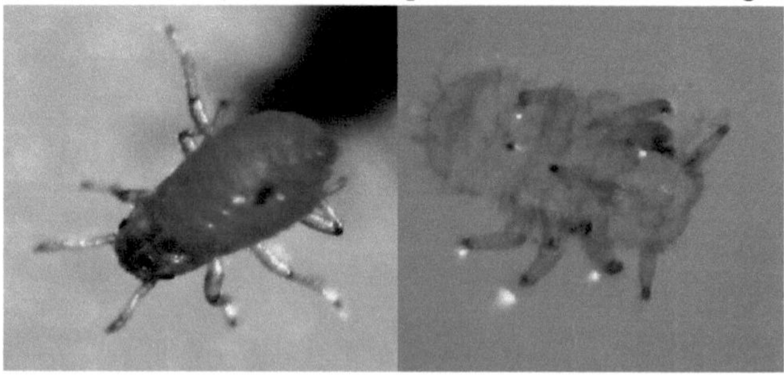

Photo19: Golden aphid (Forda riccobonii) observed under stereoscope

CHAPTER 5

DISCUSSION

5.1 - Tree dendrometry

The height of the trees reflects the quality of the environment, soil and climate, but above all the water regime, which is the major criterion in the Mediterranean climate (Quézel, 1980). At the three stations, the height of the trees in the Kheneg daya is greater than those in the Snober and Mrigha parks, despite the fact that the latter two are regularly irrigated. This difference may be linked to the age of the Kheneg stand itself, which appears to be older than the other two.Girths at 1.3 m also depend on the availability of water. They develop more rapidly if water stress is eliminated by supplementary irrigation during the growing season (Quézel and Médail, 2003). The development of the circumference depends on the development of the growth rings, which in turn reflect the quality of the climate (quantity of water) received by the plant (Roland et al., 2008). If the year is good, the ring thickness will be greater than in years of drought (Roland et al., 2008). The same applies to the number of branches observed at the three sites, a criterion that apparently varies according to age and genetic specificities from one population to another. The number of branches in a tree enables it to make better use of the environment (air humidity) and to create a microclimate under the tree itself.

5.2- Leaf biometry

Zohary (1952) used leaf morphology, especially the shape as well as the number, size and orientation of the leaflets as the first morphological character in the classification of pistachio species. These parameters seemed to be linked to a number of climatic parameters such as rainfall, average temperatures and especially minimum temperatures (El Zerey-Belaskri and Benhassaini, 2015).

a. Leaf length :

In terms of length, measurements at three sites showed a remarkable difference. The greatest lengths were reported for older pistachio trees in Snober's Garden Subjects with an average of 13.54 cm. Similar results were reported by El Zerey-Belaskri and Benhassaini (2015) in northwest Algeria. However, shorter lengths were reported for older pistachio trees in the Laghouat region (Smail Saadoun, 2014) with an average of 8.24 cm and Belhadj et al. (2008) in nine

Algerian localities including Laghouat. The same applies to subjects measured in Morocco (Abdelkader et al., 2005; Yaaqobi et al., 2009) with an average of 9.56 cm and Benabdallah (2012) in three Algerian steppe stations (Ouled Djellel, Ain Oussara and Messaad) with an average of 10.07 cm. It should be noted here that leaf growth rarely exceeds 12 cm in length for the Atlas pistachio (Khaldi and Khouja, 1995).

b. Leaf width

The leaf widths obtained in our work are similar to those reported by Smail Saadoun (2014) in Laghouat, Benabdallah (2012) in his three Algerian steppe localities, Belhadj et al. (2008) in several Algerian localities and Yaaqobi et al. (2009) in Morocco as well as El Zerey-Belaskri and Benhassaini (2015) in 16 sites in northwest Algeria.

c. Length and width of terminal leaflet

The average length and width of the terminal leaflet in the three sites showed similar values to other pistachio populations (Yaaqobi et al., 2005; Belhadj et al., 2008; Benabdallah, 2015; El Zerey-Belaskri and Benhassaini, 2015). However, Belhadj et al (2008) reported a significant difference between the measurements of the terminal leaflets at the same site, which calls into question the use of this parameter as a criterion for classifying the Atlas pistachio.

5.3- Golden aphid damage in pistachio stands: Cecidogenic aphids

Aphids have a very weakening effect on plants, producing large quantities of honeydew that encourages the development of fumagine. They transmit a large number of viruses (Zucker, 1982; Martinez, 2008; 2009). Winged aphids can travel several kilometres by passive flight with the help of the wind. According to Forrest (1987), 700 of the 4,400 aphid species described worldwide produce a gall in the course of their life cycle, in which they complete part of their life cycle. Many others deform plants t o varying degrees to provide shelter. The spread of galls on the leaves from the first sight to the appearance of the fruit. The latter was reported by Belhadj (1999) for the golden aphid in the Ain Oussara pistachio stand. This was also the case in our study area, where a very heavy attack of this insect was noted. Some pistachio trees are completely affected, and the attack does not distinguish between young and old trees. Two types of galls have been observed, one that is red in colour and slightly elongated in shape, with a low frequency, and the other that is spherical, about 1cm in diameter and sometimes more, yellow in colour, with a very high frequency on the male plants at our stations, turning brown in autumn.The results obtained show that the average aphid attack on Atlas pistachio was 165291± 0.4 per site. According to the results of the analysis, the average number of galls per plant is 7.74 galls, ranging from 0 to 80 galls, which remains low compared with those reported by Martinez (2009) in Palestine.We noted that the most attacked trees were those in the Daya de Kheneg, followed by those in the Snober garden and finally, with a low attack rate, the young pistachio stand in the Mrigha park with a large number of female trees affected compared to males. This situation is probably linked to the morphology of the trees: the largest trees (height and volume of the crown) are the most exposed to aphid attack. These insects synchronise their installation on the trees with the formation of important foliage (Burstein and Wool, 1993) while ensuring the development of the different parts of the plant species, particularly the widths and lengths of the leaves and leaflets (Zucker, 1982; Price, 1984, 1991; Whitham, 1992).The influence of station factors (irrigation, water stress, proximity of roads and buildings, etc.), microclimate and edaphism on the morphology of plant species determines the degree and density of infestation (Zucker, 1982; Martinez et al., 2005).In the daya of Kheneg, this insect is very responsive due to the location of the site near the road (Laghouat - Kheneg road). Trees growing along roads were more often parasitized by the Forda riccobonii aphid and had more galls in this disturbed environment and on artificial sites than on natural sites (Martinez and Wool, 2006). This is also the case for Snober. These attacks could halt shoot development (Martinez, 2008) and can also induce physiological factors and morphological changes in plant tissues (Tscharntke, 1989).According to Itzhak (2008), the aphid (Forda riccobonii) has a complex life cycle, forming two different galls on the host tree. The first, created by the fundatrix in spring, is a small red ball (<5 millimetres) on the main nerve of the leaflet.The second, established a few weeks later by the direct progeny of the first galls (F2), which migrate to other leaves and form an order of variable numbers of spherical red hinged chambers on the leaflet margins, and it is within these that they reproduce. The ecology of Pistacia atlantica populations and the types of galls observed on this species have been considered in the identification of the aphid species (Martinez, 2009; Álvarez et al., 2016).

CONCLUSION

Our study consists of assessing the damage caused by the golden aphid (Forda riccobonii) to Atlas pistachio trees in the Laghouat region by estimating the infestation rate of this aphid and establishing a link between the biometric parameters of the stands at the three sites and the effects of this insect. The three sites have different Pistachio stands in t e r m s   o f dendrometry and foliage; the M'righa site is the youngest compared with the other two sites. The number of galls is highest in the El Kheneg daya, followed by the Snobar garden and lastly the M'righa site. This number is closely related to several dendrometric parameters such as tree height, crown height, trunk height, leaf width and length of terminal leaflets.The results showed that female plants were more affected by the golden aphid than male plants. Our results showed that pistachio trees in the Laghouat region are exposed to attack by several aphid species. The aphid fauna recorded at three sites in the region is made up of 03 different species, namely : Forda riccobonii, Geoica utricularia and Smynthurodes betae. They all belong to the Aphidinae sub-family.There are several types of gall, red and green in colour and spherical and articulated in shape, about 1cm in diameter, which change colour with the seasons.With regard to the parasitism of these species on Pistachio trees, the results showed that the degree of infestation of Pistachio trees by aphids in the Laghouat region varies from one site to another. Daya de Kheneg has a high level of infestation, with 14 infected trees out of 25, giving a rate of 56%. The Snober public garden is moderately infested, with 8 infected trees out of 28, with a rate of 28.57%, and a low infestation in the young stand of pistachio in the M'righa amusement park; only 4 trees out of 26 were infected, with a rate of 15.38%. The Atlas pistachio must be given all the care it deserves. More research is therefore needed into the health status of this species and environmental stresses, as well as its specific ecological and physiological tolerances. It is also essential to continue research on aphids in different regions on a larger scale on pistachio, for longer periods, and to study the number of aphids in the gall, the number of chambers forming the gall, to analyse the habitat of the galls, the shape of the galls, their size and to follow the aphid development cycle inside the gall. Plant resistance (variety) to aphid behaviour. We also need to look at the different relationships between beneficials and aphids as a means of biological control.

REFERENCES

Benabdallah, F. (2012). Morphological study of the leaves and fruits of the pistachio tree of the atlas (Pistacia atlantica Desf) and valorization of the essential oils of the leaves and the oleoresin. Master's thesis, biotechnology option, Mohamed Kheider University Biskra :p37

Benhassaini, H. (2007). Phytoécologie de Pistacia atlantica Desf. subs Pitacia atlantica dans le nord-ouest Algérien article scientifique, sécheresse 2007 :p199-205

Bellefontaine, R., Petit S., Pain-Orcet M., Deleporte P., Bertault JG. (2001). Trees outside forests: towards greater consideration. Cahier FAO Conservation (Rome), : 35, 214 p.

Burstein, M. & Wool, D. (1993). Gall aphids do not select optimal galling sites (Smynthurodes; Pemphigidae). Ecological Entomology, 18: 155-164

Blackman, RL. Eastop, VF (2006). Aphids on the World's Herbaceous Plants and Shrubs. Vols 1 and 2. Wiley, Chichester and New York :1439 pp. C. Agabi, "Daya", in Gabriel Camps (1995.) . Daphnitae - Djado, Aix-en-Provence, Edisud (" Volumes"No 15).
Boudy, P. (1952). Guide du forestier en Afrique du nord. Vol 1, Edit. La Maison rustique, Paris :509p.

Capot-Rey, R. (1937). "La région des dayas", Mélanges offerts à E.-F. Gautier : p. 107-130.

Despois, J. Estorges, P.(1949). "Morphologie du plateau Arbaa", Travaux de l'I.R.S., t. XVIII, 1957, p. 23-56 and t. XX, 1961: p. 29-77.

Chaba, B., Chraa, O. AND Khichane, M. (1991). Germination, acinar morphogenesis and growth rhythms of the Atlas pistachio tree (Pistacia atlantica Desf.). Physiology of trees and shrubs in arid and semi-arid zones. Tree Study Group. Paris, France: P 465-472
Dubief, J. (1950). Evaporation and climatic coefficients in the Sahara. Ed: I. R. S., Tome VI, Algiers: 13-43pp.

Dubief, J. (1953). Essay o n surface hydrology in the Sahara. General Government of Algeria. SES Clairbois-Birmandreis. Algérie.

Dubief, J. (1959). The climate of the Sahara. Tome I, Les températures. Travaux de l'Institut de Recherche Saharienne :312 p.

Dubief, J. (1963). The climate of the Sahara. Tome II. Fascicule 1, Les précipitations. Travaux de l'Institut de Recherche Saharienne :275 p.

Eastop, VF. (1971). Deductions from the present day host plants of aphids and related insects. The Royal Entomological Society of London 6: 157-178pp.

El Zerey-Belaskri1, A. and Benhassaini, H. (2015). Morphological leaf variability in natural populations of Pistacia atlantica Desf. subsp. atlantica along climatic gradient: new features to update Pistacia atlantica subsp. atlantica key. International journal of biometeorology 60 (4): 577-589pp.

Estorges, P. (1961). Morphology of the Arbaa plateau. Trov lnst Rech Sobar; Xx : 29- 75pp.

Fraval, A (2006). Aphids. Insectes. No. 141: 3-8pp.
Godin, C, Boivin, G. (2000). Guide d'identification des pucerons dans les cultures maraîchères au Québec, Agriculture and Agri-Food Canada, 4-30pp.
Godet, J. (1988). Arbres et arbustes aux quatre saisons, Delachaux et Niestlé Paris.
Harfouche, A., Chebouti, N., Meziou and Chebouti, Y. (2005). Comportement comparé de quelques provenances algériennes de pistachier de l'Atlas introduites en réserve naturelle de Mergueb (Algérie) , forêt méditerranéenne t. XXVI, n° 2 :p135.

Iluz, D. (2011). The plant-aphid universe. Cellular Origin, Life in Extreme Habitats and Astrobiology. 16: 91 -118pp.

Khaldi, A., Khodja, M.K. (1996). Atlas pistachio (Pistacia atlantica Desf.) in North Africa: taxonomy, geographical distribution, utilization and conservation. Genetic Resources. IPGRI, Rome, Italy: 57-62pp.

Leclant, F. (1999). Les pucerons des plantes cultivées : clefs d'identification. Il cultures maraichères, INRA. Paris: 9-14pp.

Maamri, S. (2007). Etude de Pistacia atlantica de deux régions de sud Algérien : dosage des lipides, dosage des polyphénols, essais anti leishmanies, Mém. Mag. Uni. M'hamede BOUGARA Boum : 96p

Monjauze, A. (1968). Note sur la régénération du Betoum, par semis naturels dans la place d'essai de Keflafaa. Bul. Social. Natural history of North Africa.

T:56p.

Monjauze, A. (1968). Distribution and ecology of Pistacia atlantica Desf. In Algeria ; Bul. Sociol. Histoire naturelle de l'Afrique du nord. 56(2): 5-128pp.

Monjauze, A. (1980). Knowledge of Betoum (Pistacia atlantica Desf.). Revue forestière Française. Biologie et forêt. N 4 : 357-363pp.

Martinez, J. Wool, D. (2003). Differential response of trees and shrubs to browsing and pruning: the effects onPistacia growth and gall-inducing aphids. Plant Ecology 169: 285-294pp .

Martinez, J.-J Y., Mokady, O. & Wool, D. (2005). Patch size and patch quality of gallinducing aphids in a mosaic landscape in Israel. Landscape Ecology 20: 1013- 1024pp.

Martinez, J.-J.Y. & Wool, D. (2006). Sampling bias in roadsides: the case of galling aphids on Pistacia trees. Biodiversity and Conservation 15: 2109-2121pp.

Martinez, J.-J.Y. (2008). - Impact of a gall-inducing aphid on Pistacia atlantica Desf. Trees. Arthropod-Plant Interactions, N°2 :167-151pp.

Massenet, JY.(2005). Lycée forestier - Château de Mesnières - 76270 MESNIERES-EN-BRAY: 25-65pp.

Nesson, Cl. (1967). "Evolution d'un "bétoir" dans la daïa M'rara à l'ouest de l'Oued Righ" Travaux de l'I.R.S., t. XXI: 67-77pp.

Oauphin, P. (1993). - Les Galles de France - Mém. Soc. linnéenne de Bordeaux, 2, 316p, 112pl.

Ozenda, P. (1983). Flore et végétation du Sahara. 2nd ed. CNRS. Paris.624p
Robert, Y. (2008). - Aphids, vectors of virus diseases, . Revue de zoologie agricole appliquée : 34.

Simon ., J-C, Stoeckel ., S, Tagu ., D .(2010). Evolutionary and functional reproductive strategies of aphpds. C.R. Biologies: 488-496pp.

Simon, J-C. (2007). When aphids socialise. Biofutur: 13-39pp.

Quézel, P. and Médail, F. (2003). Ecology and biogeography of Mediterranean forests. ELSEVIER, Paris :11-31pp.

Riedacker, A. (1993). Physiology of trees and shrubs in arid and semi-arid zones, 489p.

Saadoun, N. (2005). Stomatal types of the genus Pistacia: Pistacia atlantica Desf. ssp. atlantica and Pistacia lentiscus L. Options méditerranéennes, series A 369-71pp.

Roland, J.-C., Roland F., El Maarouf-Bouteau H. and Bouteau F. (2008). Atlas Biologie végétale, 2. Organisation des plantes a fleurs. Ed. Dunod, Paris, 144p.

Rouvinen, S. and Kuuluvainen, T. (1997). Structure and asymmetry of tree crowns in relation to local competition in a natural mature Scots pine forest. Canadian Journal of Forest Research, vol. 27 : 890- 902pp.

Taïbi, AN. (1997). Le piémont sud du djebel Amour (Af/os saharien, A/gériel: opport de la télédéfection sate//ifaire à /'étude d'un milieu en dégradation. Thèse de doctorat nouveau régime, Université de Paris-VII, 310 p.

Wertheim, G. (1954). Studies on the biology and ecology of the gall-producing aphids of the tribe Fordini (Homoptera: Aphidoidea) in Palestine. Transactions of the Royal Entomological Society of London: 79-96pp.

Wool, D. (1995). Aphid-induced galls on Pistacia in the natural Mediterranean forest of Palestine: which, where, and how many. Israel Journal of Zoology 41: 591-600.

Wool, D. (2005). Gall-inducing aphids: biology, ecology and evolution. In: Raman, R., Schaefer, C.W. & Withers, T.M. (Eds.), Biology, Ecology, and Evolution of Gall- inducing Arthropods p. 73-132. Science Publishers Inc. Enfield, New Hampshire, UK. 817 pp.

Wool, D. & Burstein, M. (1991). A galling aphid with extra life-cycle complexity: population ecology and evolutionary considerations. Researches on Population Ecology,: 307-322pp.

Wool, D. & Bogen, R. (1999). Ecology of the gall-forming aphid, Slavum wertheimae, on Pistacia atlantica: population dynamics and differential herbivory. Journal of Zoology 45:247-260pp.

Yaaqobi, A. El Hafid ., L. and Haloui ., B. (2009). Etude biologique de Pistacia atlantica Desf. de la région orientale du Maroc, Biomatec Echo, 3 : 39-49pp.

Zohary, M. (1952). A monographical study of the genus Pistacia. Palestine

Journal of Botany, Jerusalem Series 5:187-228pp.

Zucker, W. (1982). How Aphids Choose Leaves: The Roles of Phenolics in Host Selection by a Galling Aphid. Ecology, Vol. 63, No. 4: 972-981pp.

Printed by Books on Demand GmbH, Norderstedt / Germany